# WINDOWS TELEPHONY

A Practical Guide to Designing,
Using and Developing Telephony Applications
on all Windows Operating Systems
at the Desktop and via Client-Server

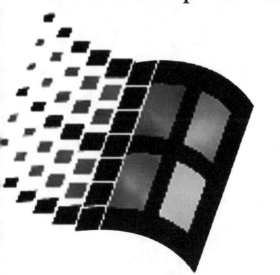

**COVERS**
Windows 95, NT, TAPI,
TAO, ECTF S.100,
CTI, IVR, FAX, MAPI,
NTFS, TCP/IP and
Telephony Networking

by *Jeffrey R. Shapiro*

**WINDOWS TELEPHONY**
**A Practical Guide to Designing, Using and Developing Telephony Applications on all Windows Operating Systems at the Desktop and via Client-Server**

A Flatiron Publishing, Inc. Book
copyright © 1996 Jeffrey Shapiro
published by Flatiron Publishing, Inc.

All rights reserved under International and Pan-American Copyright conventions, including the right to reproduce this book or portions thereof in any form whatsoever. Published in the United States by Flatiron Publishing, Inc., New York.

12 West 21 Street
New York, NY 10010
212-691-8215   Fax 212-691-1191
1-800-999-0345
1-800-LIBRARY
www.flatironpublishing.com

ISBN # 0-936648-94-5

by Jefffrey Shapiro
November, 1996

Manufactured in the United States of America


Cover design by Mara Seinfeld
Printed at Command Web, New Jersey

**Dedication**

To my wife, Kim, and my son, Kevin, with all my love.

# Acknowledgments

This book would not have been possible without the combined effort of many people.

Thanks go to the publisher, Harry Newton, and Flatiron Publishing, Inc. for the opportunity to write this work. In particular, thanks go to Christine Kern for her support and commitment to the project and to the production team at Flatiron.

Thanks go to my friend and partner, Karl Slatner, publisher of Online Business Today who made the resources (computers, networks, software and more) of Home Page Press, Inc. and himself available to me for this work.

I owe more than acknowledgments to my wife, Kim, whose support was unfaltering, albeit in a very difficult time for her. Thank you for your understanding, your patience, and your encouragement.

During the writing of this book I spent much of my time getting my "hands dirty" with some of the latest in computer telephony hardware and software products. For this opportunity I thank Rhetorex, Inc.– especially CEO Al Wokas – for the loan of their TAPI-compliant voice processing development kits for Windows 95 and Windows NT.

I would also like to thank former Marketing Director of Rhetorex and now Aculab, Inc. President, Mike Ross, for his advice and encouragement over the years.

I also received much help, in the form of a continuous stream of the Visual Voice computer telephony software toolkits, from the Computer Telephony Products Group of Artisoft, Inc (formerly Stylus Innovation, Inc.) in particular Vice President of Marketing, David Krupinsky.

Thanks also go to IDG Books Worldwide, Inc., for allowing me to use passages and several diagrams from my book Computer Telephony Strategies.

# CONTENTS

## Chapter 1: Telephony 101 .................7
- The Public Switched Telephone Network ...................7
- Devices and Instruments ................................8
- Switching Facilities and Equipment ........................10
- Local Loops ........................................10
- Telephone Circuits ...................................11
- The Basic Telephone Call ..............................12
- Understanding the Loop ...............................15
- Dialing Call Completion and Disconnect ...................18
- Depressing the Switch-hook ............................23
- Transferring a Call ..................................24
- Coverage .........................................25

## Chapter Two: Computer Telephony Basics .......27
- Computer-Telephone Integration .........................27
- The Old Era of CTI ..................................30
- Computer Telephony ................................34
- First Generation CTI .................................35
- In-band Integration ..................................35
- Out-of-band Integration ...............................36
- Call-Progress Analysis and Call-Progress Monitoring ..........37
- Call-progress Monitoring ..............................38
- Digital Integration ...................................44

## Chapter 3: Windows System Services and CT Engineering ...............46
- Mission Critical Services and Safe Systems ..................46
- The Telephony Operating Systems: Windows NT and Windows 95 ..........................50
- NT Unleashed: Symmetric-Multiprocessing Architecture (SMP) ................................52
- NT Unleashed: C2 Certified Security .....................53
- NT Unleashed: Support for Multiple Platforms ...............55
- The Win32 API ....................................55
- File Handling .....................................57
- Networking and Client/Server Telephony ...................57
- Networking: File Copy ...............................61
- Networking: Mailslots ................................61

- Networking: Named Pipes ................................62
- Networking: TCP/IP ....................................64
- Programming TCP/IP ...................................69
- Remote Procedure Calls (RPC) ..........................74
- Preemptive Multitasking and Multithreading ................75
- Services ..............................................80
- WOSA ................................................81
- TAPI .................................................82
- Building Computer Telephony Applications .................82
- Compiler Wars: "Battle of the Visual Masters" ...............84

## Chapter 4: Windows Telephony and TAPI ........94
- The Evolution of Windows .............................94
- What Does TAPI Add to Windows? ......................95
- TAPI 101: An Introduction to the Telephony API ............101
- Client/Server Telephony and TAPI 2.0 ....................107

## Chapter 5: Key TAPI Functionality ............121

**Appendix A: Windows Telephony Lexicon .......146**
**Appendix B: Computer Telephony OLE**
           **Component (OCX) Example ........190**

# Introduction

No software process is a trivial invention. No application or process is a cinch to write. Writing good software is an extremely complex discipline. The software writer has to consider the myriad different ways people will use the software, the myriad different habits in the user's life, the environment in the office, on the road, at home, two miles up in a 747, gender, age, political leaning (Apple, OS/2, UNIX . . . ) and so on. The course or sequence of events that a user performs in order to arrive at particular solutions and certain processes on the computer is almost too complex to imagine, as I am sure you know. The world of telephony adds a whole lot more complexity to those processes and, thus, the software creator's life.

The Windows operating systems—especially (and for all intents and purposes) the 32-bit operating systems—have made life a lot less frustrating for those who have been given the task of creating telephony applications for the computer. Certain processes and functions that support telephony are now available at the operating system level, and as part of operating system services. As such new abstraction layers, APIs, are being created to provide easier access to complex "low level" functionality. Telephony APIs that are beginning to enjoy wide support are coming of age.

You may on first "parse" of the TAPI documentation think that the life of the computer telephony software nerd was made a whole lot easier with the advent of the API . . . or, for that matter, TSAPI and various other APIs and toolkits floating around. But despite the new ease, it's not really all serendipity. The Win32-bit API now makes it possible to create all manner and magnitude of software to assist mankind gain control or solution over its most complex problems; however, in many respects the broader and more powerful the operating system and hardware the easier it is to become overwhelmed and to lose one's way and, yes, vision. After all its just as easy to drown in the bathtub as it is in the ocean. Or as the cliché goes, you don't need a lot of rope to hang yourself.

Take the Internet for example. It's been around for some time. Only now new applications and services are arriving that are moving it into its critical stage; into critical service. Several years ago I explored the use of TCP, the Internet's "nerve center," for connecting Windows clients to Windows telephony servers (for screen pops). I did not have to look too far. Today, so much is taking place on the Internet, and at such a rate, it's frightening. (As I write this introduction, the Miami Herald reports that telephone networks in the USA are on the verge of collapse because of Internet access mania.)

All the TCP technology in the world is right at your fingertips. Two years ago the price for a TCP/IP—Windows sockets toolkit was in the region of $25,000 (mostly through license fees and royalties that were required in advance). Today, these tools are free and there is no cost to distribute them to your users; no royalties to pay.

And while on the Internet subject . . . . To many software engineers and entrepreneurs the Internet is a monster, an unpredictable beast that will be the ruin of many fine software companies who feel the proverbial rugs are being pulled from under them. Let me share with you a scene I witnessed at the 1996 Computer Telephony Expo in Los Angeles. A computer telephony guru in charge of a successful product had setup his software feeling confident that his company had a leading edge product that would be hard for anyone to beat at the show. By the second day, he was on the verge of a nervous breakdown. Instead of interest in the product's new features, in the installed base, in questions that were so important the previous year, all he got was Internet, Internet, Internet. He came over to the Microsoft booth, desperately searching for someone who could explain what POP meant, and how HTTP could deliver voice mail and more. His most desperate question "what has any of this to do with voice mail or the telephone." "Puh-lenty."

Understanding TCP/IP and understanding how to incorporate the technology into computer telephony technology is thus essential. TCP/IP has a large part to play in any computer telephony application, not only for the purpose of "carrying" person-to-person dialog over the Internet or intranet, but process-to-process communication over a network as well. As a result I have paid particular attention to TCP/IP in this book, and I hope to expand on it in future editions and in other books.

## For Whom Windows Telephony Calls

What I hope to achieve in the few chapters ahead (at least in this first edition) is not only to steer the computer telephony engineer and fledgling Windows Telephony programmer in the right direction but to provide some basic insight into computer telephony engineering in general, and to Windows telephony specifically. But, you might ask, why Windows?

We have at our disposal several fine operating systems to control our sophisticated computers. Some, like UNIX and OS/2, were doing justice in the computer telephony industry long before "telephony" popped into the helicopter heads at Microsoft and Intel. Today, however, the Windows OS, and its API family (of which TAPI is a tiny part) has evolved to become what I consider the most viable operating system architecture on which to build computer telephony software today. It's not that the Windows operating system has something special to offer telephony gurus that no other operating system offers. UNIX, after all, had led the "hard-core" CT application field for a lot longer. Its viability has to do with scope and vision, something for which Microsoft deserves a fair amount of credit.

A computer telephony system programmer needs access to a rich API of many functions with which he or she can create processes to solve any conceivable engineering problem. The Windows (Win32) API has it. The support you get from the Windows API is unchallenged it almost every facet of computing. It has unchallenged support for messaging; it has unchallenged support for databases; it has unchallenged support for security; it has unchallenged support for telephony and serial communications (such as RAS); it has unchallenged support for operating system services (the registry and the hardware abstraction layer); it has unchallenged support for games, voice and speech processing; RAD (rapid application development and visual programming); user interfaces (UI); components; object oriented technology and more.

The Windows operating systems are now so exhaustive that you could also easily argue why point-of-sale software is better in Windows than say OS/2 or UNIX; or life-preserving medical software; or reservations systems, and so on, ad nauseum. This wide scope of support is found in one place . . . the Win32 API and the Windows SDKs. This is a collection, a library, a tome, of APIs and sub-APIs with functions so numerous and exhaustive that no single programmer alive can claim to have a complete knowledge of it.

Windows is also contending for the TCP/IP championship . . . the very service that for so many years was mainly associated with UNIX. Even traditional UNIX programming shops are switching to Windows NT for Internet services. Some claim its support for the transport protocols of the Internet is so powerful they can offer throughput (such as electronic mail transfer) at levels that far exceed the capability of the UNIX machines.

A software engineer needs access to all the above, and more, to create reliable and robust computer telephony systems. He or she also needs access to rich, robust and mature compiler platforms and development environments. For all this support he or she can rely on Windows to deliver.

At the same time I have been fortunate enough to have access and understanding to some of the other technologies under development at Microsoft, to realize that much of what we have to work with today, is virtually obsolete (only egos and political affiliations (like Sun Vs Microsoft) stand in the way of universal adaptation and adoption). Technology that will change the way we engineer and create software forever is almost here. I even predict that TAPI programming (as we do it today) will be obsolete within a year or two. As a Windows telephony guru you will get access to all the great stuff to come.

But Windows also has something that no other operating system has. Critical mass. Windows servers and clients are found in almost every office in the world. In fact, in many countries the support for Windows is running at penetration percentages that exceed levels in the USA. The USA is more Third World when it comes to Windows penetration than many Third World countries. Every business uses the telephone, and virtually every business is running Windows. This makes it the most popular client/server computing system in existence. Why would you want to support anything else?

In many respects I believe the largest audience for this book is not the computer telephony software gurus that work in the labs at Applied Voice Technology, SpeechSoft, and other specialists in computer telephony products and turnkey applications. I believe it will be more needed by the corporate developer who is tasked with the goal of building applications to solve enterprise specific problems, and to empower his or her enterprise by addressing enterprise specific goals.

## Structure of the Book

If you had a peek at Chapter 1 you may have been a little surprised to find that the chapter discusses mostly telephony, and not much Windows. It may seem desirable or sufficient to dive directly into the software engineering and the Windows Telephony API (TAPI) in a book specializing in Windows Telephony. But no matter how competent a software engineer, or programmer you are, how great a LAN or intranet guru, you should first gain a basic understanding of the world of telephony. It has been around a lot longer than Windows, PCs, Andy Grove, Billy Gates, Harry Newton . . . and me. And, if you are new to telephony, many key functions in TAPI will be lot clearer after reading up on the telephony subject first.

I find it amusing when I read articles on how to create a computer telephony product, even the "simplest" of dialers, and the authors provide little explanation of the world of telephony beyond the PC bus. What then results from the lack of knowledge and awareness are products that are not only useless, they are also annoying.

Here's an example: Consider a PC dialer (a modem and software) that shares the telephone line with your user's plain on telephone. How long do you think it will last on your user's PC if it has the habit of going off-hook to dial without detecting that a conversation is already underway? Answer: About two seconds (or about as long as it takes to dial 555-1212). Or worse: What about those voice mail systems that have never been programmed to detect and manage disconnect or hang-up or call abandonment? Have you thought about the ramifications of poor hang-up detection and handling? If you haven't or if it only slightly concerns you, you will get enough words on this subject later to make your head spin.

In Chapter 2, I have introduced computer-telephone integration (CTI). Before tackling APIs and Windows Telephony it is important to understand concepts such as inband and out of band integration, signaling, switching and more . . . how computer components and traditional telephony components talk to each other. Traditional CTI (accomplished with sweat and blood long before the advent of TAPI) will still impact on most computer telephony systems. There are bound to be many occasions where you may lose a sale, or where your project takes a serious knock, because of such issues.

Chapter 3 discusses many attributes of the Win32 API and the WOSA components, and it provides an introduction to client/server computer telephony, networking practices and TCP/IP. With all the ground work behind us we dive directly into a expose' of TAPI and Windows telephony programming in Chapter 4. Finally in Chapter 5 we study key TAPI functions and explore TAPI and Windows telephony software with some code examples and methods.

Besides TAPI there are many parts of the API that are essential to the creation of intelligent computer telephony software. Two groups on which your success may depend are messaging (MAPI-the messaging API) and the database APIs, such as the DAO SDK——the data access objects development kit. My company specializes in software development toolkits for both, and more. These kits are COM (component object model) and OLE tools—tools that make it easier for programmers to drill down to some of the most sophisticated functionality (much of which has been left untouched because of certain complexities). We have spent the past six months exposing every region of the MAPI to higher level. . . discovering along the way functions and features not found in any high level tools . . . and hardly documented. MAPI has facilities that will allow you solve some of the most complex and pressing problems that the heavy-weight computer telephony software creators try to solve. We did not have time to include them in this first edition. We hope to in future editions. But I hope some of the pointers I provide in the text will help you now.

In the meantime if you wish me to clarify something, or to try and help with a problem of some nature I would be happy to receive your email. You can reach me on CompuServe at 73353,2444 or via the Internet at 73353.2444@compuserve.com. You can also reach me (preferred) on the Internet at Nortech Software, Inc.—js@wizzkids.com. You are also invited to drop into the computer telephony lounge on the Internet at http://www.wizzkids.com/ctlounge. There should always be someone around you can ask a question or who may be able to help with a computer telephony problem.

## Chapter One

# Telephony 101

This Chapter Covers

- The Public Switched Telephone Network
- Devices and Instruments
- Switching Facilities and Equipment
- Local Loops
- Telephone Circuits
- The Basic Telephone Call
- Understanding the Loop
- Dialing, Call Completion and Disconnect
- Depressing the Switch-hook
- Transferring a Call
- Coverage

The life of the computer telephony software engineer (or Windows telephony programmer) is a lot more orderly than it was just a few years ago. Today we have some standards, APIs, against which you can code applications, with some degree of "surety" that the devices you wish to control (or hope to control and communicate with), can be controlled and will return various reliable parameters and information from the telephony environment. But to better envision how an application's logic might be coded to perform or undertake basic or advanced telephony functions, a chapter providing a basic understanding of telephony at work would be useful. This chapter does just that.

## The Public Switched Telephone Network (PSTN)

The public switched telephone network, or PSTN, is a complex system of interconnecting networks, routers, switches, multiplexers, repeaters, circuits and more. If you plan to get deeply involved in Windows Telephony, such as building a PBX on Windows NT; or you are taking on the development of an enterprise-wide voice messaging system, you will have many occasions to rub shoulders with this animal.

The PSTN in the USA is a little different from, say the PSTN, in Swaziland, which is almost exclusively built around ISDN technology. But for the most part the PSTNs of the world are pretty similar to each other. If they were like computer operating systems it would not be as easy to lift the receiver and call anywhere in the world.

Lets turn our attention away from APIs, operating systems and PC software for a short while (one chapter) and consider some simple, or rather basic, telephony. The PSTN or telephone network can be divided into the following four groups:

1. Devices and instruments
2. Switching facilities and equipment
3. Local loops
4. Telephone (trunk) circuits

## Devices and Instruments

This group consists of the telephones, line interface cards, modems (both analog modems and digital "modems"), fax machines, and other devices that you can attach to the telephone line. For the moment think only of the telephone line as the single twisted pair comprising the "plain on telephone service" still used by most businesses and homes throughout our world.

> TAPI NOTE: From the Windows telephony point of view these devices have varying capabilities and
> features that can be accessed or communicated with via the API.

Today, the PSTN or telephone network is robust enough to withstand the connection of any product you buy at your local Radio Shack or electronics store, or the telephony equipment, such as the POMS (plain old modems) that are now installed in every new computer that is shipped. In the past, the telephone companies of the world, such as Bezek in Israel and Telkom in South Africa, had nervous breakdowns if you attached anything to the line that did not have an "Approval Number" attached somewhere.

And many software developers around the world know all too well that there are still countries where inspectors will stroll up to your home and seize anything that is not "approved" by the "department" because they fear (with some merit in the less advanced regions of the world) that the network will collapse. Generally speaking, however (especially in the U.S.), if you want to connect your toaster or microwave to the telephone network, go ahead. As long as the device complies with federal, state or some other government authority overseeing telecommunications.

I remember while working to achieve approval for the Rhetorex line of voice processing boards that the approvals laboratory required an unusually large capacitor that needed to replace the ones Rhetorex installs on all its domestic products. We pondered why such a huge piece of plastic, resin, ceramic and metal was required to be attached. We finally found out that the required component had to do with the PSTN technicians being able to tell remotely if the device being used was "department approved." If not the inspectors could rock up to your shop and clamp your innocent device into irons. After we gained the approval Rhetorex engineers always referred to these capacitors as the "big tits we have to install on our beautiful boards." Put yourself in the engineers shoes, you spend many hours designing the perfect circuit board, only to have some crazy PTT (post, telegraph and telecommunications authority) somewhere in the world rip out your components for something unsightly, costly and messy.

Getting back to the phones: Multi-line telephones also fall into this group of line devices an instruments. These can be full-blown digital telephones (that go for prices around $300 and up) or analog feature phones that carry additional wires for sending and receiving switch information and call control data. These phones are often known as multi-button phones.

TAPI NOTE: Various revisions of the TAPI documentation reference devices and instruments as follows: The line device is the physical device that is attached to the telephone line. These devices include fax machines or fax boards, data modems, or digital interface cards. It is important to note that the API reference highlights the fact that the device does not need to be physically connected to a computer. This is important: To the definition of line device, or instrument (telephone-techie term), we can also add the PBX or switch (via the interface ports). To the CIA, FBI, KGusedtoBe, MI5 and 6, Mossad, Steven Seagal, 007, et al; a line device also refers to equipment that can tap in on a telephone line and record conversations, and something with which to strangle the enemy.

But for basic telephony the line devices with which we are concerned support numerous telephonic capabilities, mainly to assist two or more humans partake in a real-time voice conversation over a certain distance. Your (TAPI compliant) application is able to send or receive information to or from a telephone network accessing the services of the line device. The line device supports one or more homogeneous channels used to establish calls.

The API refers to the line device as "the logical representation" of the actual device. It also makes a point of dismissing the termination scenario at the end of the telephone line, which is valid as far as the remote equipment is concerned. You should not concern yourself with what hardware lies beyond the "fuzzy" picture of the "end of the telephone line." But to develop telephony and CT applications, no matter interactive and non-interactive you will need to concern yourself with how to make sense of the information and data, and analog signals, that are returned from the line and "local line device," and, of course, the data and information your application returns and the actions it takes to adequately respond.

The TAPI highlights a distinction between "basic" telephony and "extended Telephony." The minimum requirement (logically) of any application that can boast telephony capabilities ala TAPI is that the compliant line device supports all of the basic telephony functions (setting up and tearing down calls). Should you wish an application to use the extended capabilities it must first determine the line device's capabilities. The TAPI function call for this information is the lineGetDevCaps; and we will discuss it further in Chapter 5.

## Switching Facilities and Equipment

This group represents the switches and computers that connect the line devices and instruments to each other. Rarely do you have situations in which two line devices or instruments partake in a conversation (isochronously or asynchronously) without the data begin repeated, switched or rerouted via numerous circuits. These switches and equipment are housed in an elaborate and complex network of central offices (COs) around the world. Companies such as Northern Telecom, AT&T, and Siemens make this equipment.

Note: If you are wondering if the traditional telephony manufacturers will ever adopt the likes of Windows NT in central offices, I have added some thought-provoking discussion on that subject in the next chapter.

## Local Loops

The local loops are the wires that connect your phones to the COs. The following section in this chapter talks more about COs and the transmission media they use.

## Telephone Circuits (Trunks)

This group comprises the lines and other transmission media (such as fiber optics) that connect the central offices and telephone exchanges (a European term) of the world. The term telephone trunk is slowly becoming obsolete. While you may have heard the term POTS "Plain Old Telephone Service" frequently mentioned, especially in TAPI documentation, POTS calls are now mostly transmitted digitally between the telephone circuits, which typically carry a large number of channels (of conversations). These digital circuits can carry one telephone conversation between two villages in a rural area, or they can carry thousands of calls between cities.

Atlanta, in the USA, has digital circuits almost entirely composed of fiber optics. In South Africa, the world's longest fiber optic link connects the city of Cape Town with Pretoria, a span stretching more than a thousand miles, and carrying more than 600 channels. Within the "loop," voice is mostly carried by analog lines consisting of twisted pairs of copper wire. Data from most computers has to first be converted to analog form by modems. The receiving modems "demodulate" the data back to digital form to feed it to the receiving computers digital bus.

The circuits nowadays employ a variety of transmission media, such as copper wires (still the most common transmission medium), coaxial cables, microwave radio, and fiber optics. It is interesting to note that while the world waits for the telcos of our planet to lay the glass necessary to transmit data at high speeds, or to install equipment to carry digital information over the loops, some companies are refining technology to provide high-bandwidth capability over typically analog transmission media. One such company is PairGain Technologies, Inc. The product they are touting is called digital subscriber line (xDSL) technology. It works over existing copper-wire telephone lines.

The technology is being made available as two services, HDSL and ADSL. HDSL provides bi-directional high-bandwidth DSL technology for equal high-speed up- and down-stream transmissions. ADSL (asymmetric DSL) will provide a high-bandwidth downstream transmission. The latter will be ideal for Internet users who spend the majority of the time online downloading information (server-to-client centric). The former is ideal for corporate intranets where machines typically behave as both clients and servers.

This bandwidth expanding technology is being rolled out at between $35 and $100 a month for service, which compares very favorably against ISDN rates. The big deal, however, is that xDSL works over existing switching and routing mechanisms already in place on the Internet and the PSTN (which is the Internet on-ramp for millions of potential cyber-users).

## PairGain's Megabit Modem

PairGain's Megabit Modem 768 delivers 768 Kbps full-duplex data over ordinary copper wire telephone lines. The modem also features a 10BaseT Ethernet port for connection to a subscriber's personal computer, and a network port for a High-bit-rate Digital Subscriber Line (HDSL) connection to the local central office.

Features:
- Fast, 768 Kbps modem boosts desktop performance
- Six times faster than ISDN, 26 times faster than a 28.8 analog modem
- Single pair 768 Kbps full duplex HDSL connection
- Versatile 10BaseT Ethernet port with intelligent bridging
- Unprecedented performance over existing copper infrastructure

PairGain's HDSL transmission technology offers key competitive advantages like full-duplex 768 Kbps performance at fiber optic quality (10-10 BER) over a single copper pair, repeaterless connectivity up to five miles, and speeds six times faster than ISDN and 26 times faster than a 28.8 analog modem.

## The Basic Telephone Call

Wherever a caller is stationed in the world his or her telephone is either connected to a private switch, or to a bigger public switch, the likes of which are owned and operated by the Regional Bell Operating Companies (RBOCs) in the USA, such as BellSouth. When we want to call someone, typically we just pick up the handset and wait for a dial tone. What could be easier?

> **Weird Science**
>
> Even with all the modern technology and Computer Telephony wonders of the world, people will always reach for the simple telephone as a means of contacting someone, to establish a call, before using any complex setup involved in dialing from a computer. This is very important to understand in creating so-called dialers.
>
> To illustrate the point I recently read several examples that explore the TAPI address capability and address resolving functions. The authors went to great pains to show how a TAPI application can delete prefix information (such as area codes) from a phone number when the user checked into a hotel and attempted to call a local number. (The number would usually be a long distance number dialed from the user's office.) Well, most people I know who check into hotels do not go to the pain necessary to connect a computer to the hotel switch to make a local call. Once they get the information from their contact manager they simply pick up the telephone and "let their fingers do the dialing." However, a dialer that lets you easily override stored telephone numbers and switch access codes could be considered to be an assistant, rather than hamper such a situation.
>
> Don't get me wrong. I am not saying that some TAPI functions (or any other technology code for that matter) have ridiculous or frivolous application. There will be many situations in which you will thank Heaven for the API. But I am saying that most people do not require computer assistance to make a simple call; and computer telephony applications succeed when they empower the users and make their lives easier, not complicate them.

One of the electronic marvels that determines this for the computer is the DSP chip, which stands for Digital Signal Processing Chip (see the DSP entry in the Lexicon).

Simple Modems and cheap dialing devices do not have this analyses and signal frequency detecting and determining capability. The wonderful device known as the voice processing card does. Voice processing cards are loaded with all sorts of fancy signal processing and converting electronics. They dial (they excel at this task) and are the key components in voice message and predictive dialing applications.

Detection of dial tone is one of the most important features to program in Windows telephony applications. Take the following scenario: Your user will most likely be connected to the modem or dialer to the same telephone line as the telephone set. This is easily accomplished with a Y-connector (see Figure 1-1). A simple dialer function or program, especially one that commands a modem with AT strings (see the AT-Dialer example source code in Chapter 5), usually has no information that a conversation is in progress on the line when it intrudes into the conversation in order to make a call. So when it "thinks" it has an all clear to dial away it instead blows away a couple of eardrums. Using TAPI's "call-state event notifications" aids in the simple detection of dial tone (see Chapter 5).

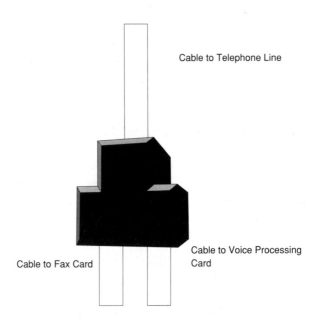

**Figure 1-1:** A Y-connector

Now let's go behind the scenes. Making a call happen is not as simple as it looks. The age old course of events is as follows: Your office extension is 414. When you want to call your assistant on extension 430, you pick up the handset and wait for the PBX or switch dial tone (my company's Centrex service works identical to the fashion I am describing here). Your phone is attached to a private branch exchange (PBX) or switch and the dial tone you hear comes from your office's PBX. It's not the dial tone that comes from the switch equipment out on the public switched telephone network (PSTN).

Calling from home is no different. The dial tone you hear is the tone sent by the telephone company. But if you are dialing from the office or behind a local or private switch you or your application will need to first dial the exchange access digit. This action puts the user "onto" the PSTN. In the U.S. the access digit is the 9 you dial before the area code and telephone number (in hotels in the US it's usually the 8 digit). (Telephony applications need to be cognizant of the prefix digits, and you will need to code the applications to meet the respective call conditions.)

But before getting too deep into the prefix subject, let's examine the process that takes place when you lift the handset. The switch CPU recognized that you took the telephone "off hook." (In telephony jargon it means that you requested a service from the system.) In techie jargon, this is the point where the user's telephone device sends the switch's CPU a so-called "supervisory" signal. Your phone or dialer generates this supervisory signal by "seizing" the phone line either through a loop start or a ground start signal to the PBX. In the digital world this so-called supervisory signal is sent along an external channel (out-of-band) to the channel that will carry voice.

## The Loop

A plain and simple telephone conversation between two people can be achieved (and is) with the two wires that make up the twisted pair. The wires that connect the telephone to the switch are known as the loop when a bridge is place across these wires they form a circuit (see Figure 1-2).

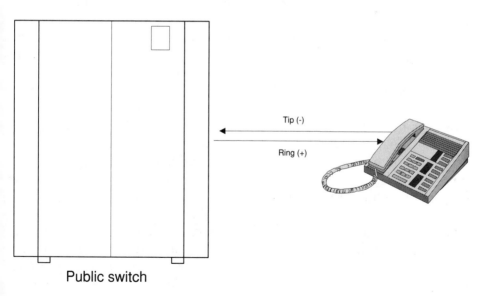

Figure 1-2: The local loop

You need current to carry the signal—your conversation—through the loop. In analog transmission the telephone company provides this current in the form of DC voltage across the wires. This voltage is commonly known as loop current. The above situation is essentially your plain old telephone service (POTS) exposed.

When you take the telephone off-hook (that is, when you remove the handset from the cradle), the lever in the cradle, known as the switch-hook, springs back to complete the circuit (it bridges the loop wires).The loop wire that carries the signal consists of two copper strands twisted (the twisted pair). This wire is generally known as unshielded twisted pair (UTP). Twisting the wires protects the signals carried by the current against electrical interference from other electronic devices.

The twisted pair is basically all you need to make a telephone call. As mentioned earlier, you have two ways to seize a line. With the loop start method of seizing a line, you start the line basically by bridging the end of the wires. This action takes place when you lift the handset out of the cradle and release the switch-hook. The negative end is called the tip, and the positive end is called the ring (see Figure 1-2).

The other method of starting a line is known as ground start. With the ground start method, one of the wires is briefly grounded to an earth connection. Most PBXs work off ground start lines, especially in the USA. Telephones and small switches mostly work off loop start lines. Callers and PBX customers usually remain ignorant of the line start method. TAPI applications do not need to factor in the methods by which a service is obtained from the line. Ground start is in any event handled by the electronics and circuitry of a device such as a PBX (and this becomes a service provider's problem). If you are setting out to design a Windows NT PBX, however, you will have to consider line starting. Switching card and PBX electronics toolkits and manufacturer APIs provide the engineer with the appropriate functions for starting lines. This is not a function of the Windows telephony API.

When your application requests telephone service, the switch or exchange acknowledges the request by providing the line device with dial tone. It then makes a connection between your telephone set and a digit receiver, which waits for you to dial. When you hear the dial tone, you know that you can now dial the digits 414 (digits are those DTMF tones that represent the numbers you dial, not your finger). Simple modems that contain no signal processing circuitry cannot detect dial tone, and the default action will be to program device dial when off-hook state is achieved, but as mentioned before this is not very professional. There are several species of PBXs that can send no dial tone to the user or line device. The dial tone was originally intended to tell humans when to dial. My company encountered this situation over the years—in some cases it was a switch feature, and in other cases it was a switch defect. So we simply programmed an option to dial as soon as the device went off-hook.

Dial-tone cadences and frequencies are not consistent throughout the PBX or telephone industry. The frequency and cadence varies remarkably from device to device, to such an extent that a computer telephony or Windows telephony system can collapse from confusion. Humans have been taught to dial when they hear a dial tone, any dial tone. Ask yourself if you automatically start dialing when you don't hear a dial tone. You probably conclude that the PBX is out of commission, has temporarily crashed, or that the phone company cut you off because you forgot to pay the telephone bills.

Sometimes, instead of dial tone, you hear a reorder tone which means that the telephone circuits are blocked. Some modems boasting dial-tone "detection" features are designed to suspend dialing when they detect a dial tone other than the one they are programmed to dial on. When the central office or PBX uses a different cadence to signal that there is voice mail on the CO's voice mail system many modems go on strike at the insult and refuse to oblige. TAPI, as we will see in the forthcoming chapters, gives you the power to force the line device to dial no matter what tone (or no tone) it hears.

## Dialing

At this point, it may be a good idea to examine the two technologies used in the dialing process. More than likely you are familiar with the DTMF tones, commonly referred to as touch tone in the United States. Pulse or rotary-dial telephones are still the norm in many parts of the world, even in some parts of the USA. In fact, the total world population of pulse- or rotary-phone users is about 85 percent. In the U.S., about 35 percent of the population still uses rotary phones. So, what's the difference between pulse- and tone-dial technology? Before the advent of touch tone, a caller advised the switch of the number he or she wanted to call by opening and closing the loop current provided by the switch, thus switching the current on and off. The techie term for this process is called loop current disconnect, or, more commonly, loop disconnect.

This process of dialing generates pulses of loop current (hence the name). In other words, to indicate "1," your phone stops the DC current flow by breaking the circuit once. To dial "9," the phone switches the current on and off nine times; for 0 (zero) the phone interrupts the current ten times. You can also use the switch-hook to break the circuit as well. But if you keep your finger down for a millisecond too long, you risk permanently disconnecting the call.

The rotary dial was invented to "automate" this circuit making and breaking exercise. The frequency of the pulses varies from PSTN to PSTN, although it is consistent throughout a country's entire telephone network. The PPS of a country can be anywhere from around nine pulses per second (pps) to around 22pps. In South Africa it's around 11pps. Israel lets you dial at 22pps. As a result dialing is a lot faster in Israel than in southern Africa,

As the technology matured, the rotary dial was replaced by a push-button keypad. But the only difference is in the fancier electronics and not in loop pulse make and break. It is even common to be working with a fully computerized digital PBX only to discover that the telephone service beyond the enterprise is loop disconnect or rotary dial, as is the case almost everywhere in Portugal. I have worked on many sites connected to these older exchanges (in cities as modern and technology savvy as London), and I have no doubt that many still exist in the world out there, so be prepared for them. All computer telephony applications or dialers should allow the user to switch between pulse dialing and tone dialing (we look at some of this code in Chapter 5).

When the PBX is connected to a pulse-dial exchange, the telephone instruments of the PBX may be rotary or push-button loop disconnect dialers. There is a valid reason for this: To dial an outside telephone number, you have to transmit the loop disconnect representations of the numbers, because the telephone exchange will only accept that type of signal. Later, as users demanded touch tone for faster internal dialing, the PBX manufacturers had to install tone-to-pulse conversion circuitry in the PBX to make the external call to the exchange. The user could merrily dial with touch tones and enjoy the speed of tone dialing, but to make an outside call, the PBX has to trap the tones and convert them into pulses for the benefit of the old, clunky, rotary-dial exchanges beyond the wall of the enterprise.

## Call Completion and Disconnect

When you hear dial tone the PBX has commissioned a digit receiver, and you can dial your number. As soon as the PBX detects the first digit, it discontinues the dial tone, and the digit receiver collects the string of digits . . . in our example this number is 4-1-4. The PBX then checks the number 414 for validity. In other words, it checks to see whether 414 is an actual extension and it verifies whether the extension (or subscriber phone) is in service or has some condition attached to it.

The verification process is important because the extension may not exist or perhaps it can only be dialed under certain conditions. Maybe the extension you dialed is a "phantom" extension, or perhaps it's an extension pool that represents a small hunt group. It could also be a service and not an extension at all, like a night or after hours bell. In such cases, you may not be able to make the call, and the PBX will send back so-called reorder tone. Reorder tone basically tells you that the "extension not available; please reorder the dial tone."

While in the U.S., this tone is called the reorder tone, elsewhere in the world, it's referred to as the service unavailable tone. This tone often sounds like a busy tone, only it's much faster. API jargon refers to the reorder tone often as "fast busy".

The fast busy signal or reorder tone relates to a concept in telephony known as blocking. Blocking is a condition that occurs when a telephone network is unable to connect two parties because all available paths between them are in use. We talk of both blocking and non-blocking networks. The public switched telephone network is a blocking network. Dedicated data networks are built around non-blocking architectures (mostly because acknowledgment packets need to be returned). (You'll run into this term in several parts of this book. See also the Lexicon.)

In the past, blocking on a voice communications network was considered acceptable because telephone conversations were generally short. However, that's not always the case anymore. With the surge in telephone traffic and the incredible popularity of the Internet and on-line communications, a level of blocking now exists on our telephone network that our telephony forefathers would never have dreamed possible. For example: I hate surfing the World Wide Web hunting for information. It is a tedious effort and extremely time-consuming. Instead I use an intelligent search agent called Go-Get-It. I use this software extensively to find information for my books and research. If I let it loose on the Internet it will stay on line for the whole night faithfully bringing me back documents (it found me more than 500 related to TAPI in under 15 minutes; and you should drop in on http://www.hpp.com to get a copy of this software). So, while I am connected to the Internet no one can reach me. Then there's the case of the nerd who promoted his Web site by spreading the word that his girlfriend was going to reveal her flesh on his home page. That exercise almost shut down all the entire telecommunications network.

There are smart PBXs that can send you a digitized voice message stored on the PBX itself instead of reorder tone to indicate a problem. The French equipment have these nifty features. You might hear words like "extension unavailable, please confirm the number and try again" in a petite French accent. The central office switch in the US does this too, telling you that the "number has been disconnected." In the U.S., if you hear this message, more than likely it means that the person you are calling has not paid the telephone bill. (I don't know of anyone in the U.S. who temporarily disconnects a service voluntarily; after all, many of us depend on our voice mail if we're not home.) However, this rather embarrassing announcement gets the job done: it gets bills paid. Elsewhere in the world most telephone exchanges just return the ring-back signal though. Here's a funny story (kind of): I was at a friend when his telephone died. There was nothing on the line . . . except silence. When we called his number from a cell phone the BellSouth recording said "The number blah, blah, blah is being checked for trouble." When we called the CO to investigate, the operator said: "The trouble is that the account is in arrears."

Now back to our simple telephony course. If extension 414 is an active extension and you have authority to dial it from your extension, the PBX will send you ringing tones (or ring-back signals ) if the extension is not busy. If the extension is in use, the PBX will send a busy tone. But if the party on 414 answers, the switch sets up a connection between 414 and 430. The process of establishing the connection is so fast that it seems as if you were directly connected to the person's telephone in the first place; that they had only to lift the handset. But during the time you hear ringing, no such connection existed. Figure 1-3 presents a graphical example of this process.

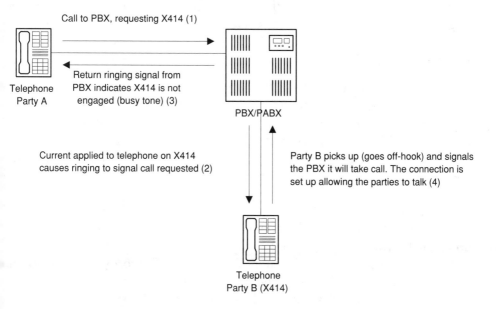

**Figure 1-3:** This illustration shows that the calling party's telephone signals to the switch that it wants to connect to the telephone on extension 414. The PBX checks its circuitry and if the telephone on 414 is free it sends current to the instrument and causes it to ring. It then sends a ringing signal to the caller to inform the device or caller that the telephone on extension 414 is ringing. If the target telephone is in use a busy signal is returned to the caller. When the called party goes off-hook (picks-up) the PBX connects the parties and a call is established. You go through the same proces if you want to make a call across the public telephone network. When you lift the handset on the telephone connected to a PBX, you first get local dial tone; that is, the dial tone provided by your company's PBX. But you don't want to dial an extension there, so to get the PSTN dial tone, you need to access the public switch through your local private switch. To do this, you press the access digit, which is typically 0 (zero) in Europe and 8 or 9 in North America.

By pressing the access digit, you tell the PBX to connect you to an external line. If all the outside lines are busy, the PBX sends you a reorder tone. If a line is available, you hear the dial tone from the PSTN.

If you are on the receiving end of the call, you know that someone is trying to call you because the PBX or PSTN prompts the telephone set to ring. It does this task by sending A/C voltage along the line to the telephone. The electronics of the telephone device can detect this voltage, which is what causes the bell to ring. Of course, the caller does not hear this ringing; the caller hears the ring-back signal . . . a confirmation that the remote set is ringing. When the called party answers, the ringing process terminates.

## Disconnecting

After one of the parties replaces the handset, that party's phone is said to be back on hook. On hook means that the switch-hook is depressed and the circuit is broken. The terminated circuit cuts the loop current and thus the connection. Although either party can disconnect the call in most cases, many exchanges require the calling party to disconnect before tearing down the connection and dropping current.

Almost everywhere else in the world, at least in places where I have worked, the central office informs the connected device of the other party's abandonment by transmitting a control signal. This tone is often called a disconnect or hang-up tone. When I worked with Rhetorex, Inc. in South Africa to gain approval for the Rhetorex line of voice processing boards one of the first requirements was to demonstrate the ability of the board to detect hang-up or disconnect tone. I informed the tester that the Rhetorex board had sophisticated DSP technology on it and could likely detect a fly buzzing around the lab. We proceeded to show the board detecting the signals on the telephone and displaying the correct frequencies on the monitor. In order to demonstrate hang-up detection we had only to identify the signal transmitted on disconnect and code the application to print the words "hang-up detected" to the screen.

The tester from the PTT's telephony labs said if the board fails this test it loses approval, which meant we could not sell the Rhetorex voice boards to clients in South Africa (well we could sell them, but we did not want to risk the threat of incarceration on Robin Island prison, which makes Alcatraz look like a Club Mediterranean resort.) Needless to day, the Rhetorex equipment passed this test 110 per cent. When the inspector saw the words pop-up on the monitor he passed the board. We could have popped up the words on the detection of any signal and the board would have passed.

Some countries have laws that govern calling party disconnect. In North America, the central office alerts the PBX or remote device to call termination or abandonment with a drop in the loop current (this is known as loop-current off transition). This control signaling is known as in-band signaling (discussed later).

In North America, many PBXs do not extend the disconnect information (the drop in the loop current, for example) to the line device at the extension (usually the computer telephony system). Because you do not get a tone there, the only way you can determine that the call is over is from silence. Humans may be able to quickly deduce that silence means the call is over, but a modem, dialer, or voice processing card attached to the analog extension will have to work on a time-out scheme to recognize that the line has essentially "died." In other words you have to program the device to analyze pure silence

In Europe and South Africa, the PTT (Post, Telephone, and Telegraph administration) approval process requires the computer telephony systems to release the line when the person called hangs up on them. The computer telephony system must be able to detect a called-party disconnect signal in order to release the line. If it does not detect the disconnect, then when the called party tries to make a call, the computer telephony system will still be holding the line. This situation can lead to a serious problem if the called party wants to make an emergency call and finds the line still open. Telemarketing machines can cause this problem and laws have been introduced world wide to prevent a disaster. It is thus vital you factor in the engineering resources to reliably and rapidly determine disconnect and hang-up (we will talk more on this subject later).

## Depressing the Switch-hook

The act of signaling the switch for service from a telephone or line device is flashing (and this does not mean taking off your clothes at the local bus stop). As mentioned earlier this is achieved by depressing the switch-hook, or hitting a recall button (which does the same thing). You need to depress the switch-hook long enough to send a signal to the switch, but not long enough to disconnect the call (unless you intend to disconnect). This flashing is the reason we often refer to the switch-hook as the flash-hook. We flash the switch-hook to put someone on hold and take them off hold.

The duration of the flash varies from switch to switch. The flash time is measured in milliseconds. If the duration of the flash is too short, the switch will not receive the signal. If the duration of the flash is too long, you may disconnect the caller or terminate the call; it's a rather delicate process.

Flashing the switch-hook for service is also known as recalling the switch. To prevent an accidental disconnect, many telephones include a recall button. When you depress the button you flash for the exact duration needed to recall the switch for service.

When you flash the switch-hook, the PBX places the passive party on hold while the controlling party (which can be also be a device such as a voice processing card) dials a service access code. This code, usually tone dialed, can signal the PBX to transfer the passive party, recall the passive party, or engage additional parties in a conference. The process is virtually the same when working with the central office switch. It's essentially how Centrex works.

## Transferring a Call

There are two ways you can transfer a call: transfer blind or transfer-on-consultation.

**Transfer Blind**

The ability to transfer calls is key to any computer telephony or switching application. TAPI, in fact all APIs, provide the necessary interface to achieving intelligent transferring, conferencing and bridging of calls (we discuss these in Chapter 2 and Chapter 5). When you manually transfer blind, you simply put the caller on hold, dial an extension number, and hang up the receiver (depress the switch-hook). Any dialer, even the simplest modem, can be programmed to do this (via simple in-band signaling). The caller you have on hold is transferred to the extension, and the receiving party has no prior knowledge of the call. The phone just rings at his or her desk and its gets answered . . . or it doesn't. In this situation, the transferring party does not care or need to know about who will answer or where the call finally ends up; the call could end up at a human, an ACD system, a computer telephony system, or it could carry on ringing until the poor person holding on dies of old age.

**Transfer-on-consultation**

Most calls are transferred on consultation (or on verification). When you transfer a call on-consultation, you wait for the called party to answer first while the caller is still on hold. You could listen for pick up and then drop your end to connect the parties, or you could consult with the called party before connecting him or her to the caller you have on hold. Why else would you consult the called party? You may want to check that the extension is the correct one (the kitchen might be on 200 and the boardroom might be on 280, and you don't want to find yourself connecting the meat pie supplier to the CEO). Other reason would be that you may need to identify the caller to the called party ("pst, it's da Don on the line, says you owe him 10 big ones. Ya want ta talk to him now, or should I say ya died in ya sleep last night.) CT and computer-based switching systems need to do the same. You need to provide the necessary logic and intelligence to make sure these electronic marvels do not make a muck of your client's operation. They should be able to transfer blind and on-consultation, perform switch control via in-band data, out-of-band data and via computer network and operating system data.

**Coverage**

Calls that are not answered after a given time (the magic number is three rings) need to be diverted. The last thing your client needs is that callers abandon after the umpteenth ring. This diversion or forwarding process is known as coverage in North America. In the rest of the world, the most common term is call-forwarding.

Coverage or call forwarding can be activated on all inbound calls to a port or extension, it makes no difference whether the call was dialed directly to the station as a direct inward dial call (DID) or connected by an operator or a computer telephony system or ACD system. Every enterprise will have a policy and a set of rules governing the coverage process. For example: Calls could be 1) diverted back to a live operator for a decision; or 2) diverted to another extension; or 3) diverted to a voice mail or an automated attendant.

Callers enable voice mail facilities at their extensions by sending control signals to the switch, telling it to divert any calls that are not answered after a certain amount of rings to an extension used by a CT system port. If you're programming a PBX or ACD system, coverage will be one of those essential services you will have to supply. If you are programming a voice messaging system this will be a tricky service to cater to. You will need to know who the call was originally intended for.

Coverage is common on domestic telephones in the USA. You can request coverage or call forwarding service from your local RBOC, instructing it to install a diversion service on your telephone line. The call can be diverted to any other telephone in the nation or even to an international location. Naturally, the telephone company will charge you accordingly for the diversion; the caller only has to pay for the original leg. If the diversion is long distance, then you (the called party) will have to pay the long-distance charges for the benefit of the coverage (and it can be really high, as much as 22 cents a minute).

## Summary

This chapter presented you with a simple introduction to the anatomy of a telephone call. With this basic understanding you will be able to better focus or fix your mind on how a computer telephony application works with the telephony environment beyond the confines of the application's process space. The chapter is appropriately titled Telephony 101, and is thus a basic introduction. Most city libraries carry good books on the science of telephony and telecommunications. You should consult them for further information on telephony (especially digital telephony). From here on you may wish to explore the workings of switches and PBX systems. One of the best books on the subject of telephone devices and switching systems is James Martin's Telecommunications and the Computer, published by Prentice Hall. Harry Newton's dictionary, *Newton's Telecom Dictionary*, published by Flatiron Publishing, also has loads of information on PBX and switching systems and technology.

## Chapter Two
# Computer Telephony Basics

This Chapter Covers

- Computer-Telephone Integration
- The Old Era of CTI
- Computer Telephony
- First Generation CTI
- In-band Integration
- Out-of-band Integration
- Call-Progress Analysis and Call-Progress Monitoring
- Call-progress Monitoring
- Digital Integration

All modern PBXs and telephone equipment incorporate some form of computer control or processing. In addition to these smart, sophisticated, (yet often hard to use, and paradoxically dumb) PBX systems controlled by a CPU, you can program a large variety of line devices to integrate with the services of a PBX. The objective of this software engineering is to achieve a form or computer control and enhancement over the telephone call, for the benefit of the user. This integration, or control, enables you, the software engineer to use PC programming techniques to automate the telecommunications process in a given organization or service entity. This is the essence of computer telephony, and we devote this chapter to the subject.

## Computer-Telephone Integration

From an engineering perspective, the objective of CTI essentially is to provide two classes of service:

- Computer control over the call and the ability to take and make a call
- Human-computer dialog over the telephone

The call-control side empowers the enterprise by liberating the human from the call answering, routing, filtering, and dialing process. The human-computer dialog side obviates the need to have two or more humans involved in a telephone conversation. Figure 2-1 summarizes the CTI objectives.

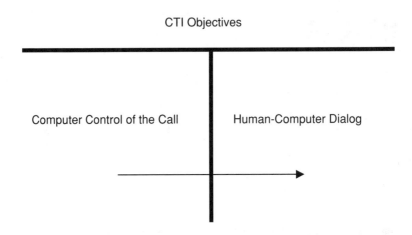

**Figure 2-1:** CTI objectives are: 1) Computer control of the call, and 2) human-computer dialog. (The arrow indicates that division 1 is the essential service.)

If all you are required to program are routines to route and process calls to agents or representatives, or automatically dial out to colleagues external to your organization, then you do not really need a computer telephony system that includes the capability to have a dialog with a caller. You could design the human-computer dialog side as a separate process or interface (even a thread of execution) if necessary. A division 1 only system is likely to be an ACD, a predictive dialing service, a personal dialer (like a PC modem or a dialer device such as the PATI from Comdial), or a sequencing system that (at the most) plays a digital recording requesting the caller stay on hold. The other mainstay computer telephony services, such as voice mail, require you to design in both services, as depicted in Figure 2-1, to program both call processing and the ability of the users to have meaningful human-computer dialog.

It is important to note here that in the past decade, computer telephony systems did not provide a user interface at the user's desktop PC. The only interface was via the telephone (and it continues to be the most widely used). So you had two interfaces: one for the owner of a mailbox or service, and one for the customers or callers. Today, the desktop represents the third interface into the world of computer telephony. This is what makes

Windows Telephony so exciting. You can get really creative now with the programming and software engineering resources at your disposal.

A word of warning. The telephone user interface or TUI is the essential component of a voice messaging or computer telephonist (automated attendant) service. Yes, you *will* have to create "windows" to your Windows telephony system on the plain old telephone. You can leave out the desktop part, the DUI, but you can't leave out the TUI. This may seem ironic because the soul of Windows is its user interface on the monitor, and the API that gives life to the interface is TAPI. You should consider splitting your team, at least, into two parts: One part, or team, to deal with "engines" or the low-level services, APIs, libraries and more; and one part to deal with the user interfaces. It is rare to find a software engineer that excels in both realms. And it is also important to remember that you will need to go below TAPI and interface to telephony equipment at lower levels because TAPI does not cater to all equipment processing.

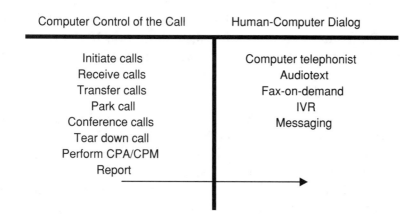

**Figure 2-2:** The two divisions in CTI objectives enable several services.

I will devote much of the rest of this chapter to the enabling technology of the first division of CTI. To qualify as a computer telephony system, your product first must, for myriad reasons, be able to automatically and manually perform one of the following functions (in no particular order):

- Initiate or set up calls
- Receive calls
- Transfer calls
- Park calls

- Conference calls
- Tear down calls and conferences
- Perform call-progress analysis (CPA)
- Perform call-progress monitoring (CPM)
- Provide real-time reporting and event logging on the aforementioned functions

Secondly, although not as important, the system should be able to perform these functions one call at time or concurrently.

APIs such as TAPI and ECTF S.100 (see Lexicon), provide the necessary "hooks" to achieving these objectives via operating system services and against standards (see Chapter 3). You could, of course, create an application that is all TAPI, or TSAPI, but in my book that would not be a product that would command great support as a commercial application. You see, most of the world's PBXs and telephony systems are NOT TAPI compliant and do not provide for interfacing with any HAPPY API. You can only integrate computers with such equipment via the long established in-band and digital (such as SMDI) signaling and data feeds. If all your app talks is TAPI, for most customers it will be still born. If you come to my summer CTI camp, you'll need to bring with your oscilloscope.

## The Old Era of CTI

DOS based telephony systems had to emulate multi-tasking software with fancy delicate state-machine executables, juggling tasks like clowns on rolling barrels in a circus act. Later moves to OS/2 and UNIX, true multi-tasking and multi-threaded operating systems, had software developers figuring they could fire up a new state machine for every port on their telephony system and that would be that. But it's not as simple as it sounds. Events such as database reading (for myriad reasons) and writing, saving voice files to disk, transferring callers, answering other ports and event tracking and analyses all have to happen concurrently. And all the tasks also have to be managed so that they do not "crash into each other." Only in the last year, since Windows 95 was released and the telephony drivers and TAPI finally made it out under 32-bit was it possible to do some serious computer telephony, especially under Windows.

Windows 3.X, may she rest in peace, was not a true multitasking operating system. Although you can run software applications concurrently in Windows 3.X the real problem lies in the underlying MS-DOS 16-bit architecture. I think almost every computer or technology hack has written about the limitation of DOS and Windows 3.X., so I am not going to delve into that subject here. What I will say is that for the large part you can dismiss being able to create anything that can be scaled up past four ports (reliably), eight at the most (if carry liability insurance). The problem lies in the memory limitations of Windows 3.X.. And Windows 3.XX and earlier was not an operating system. It was a fancy graphical user interface attached to a fickle interrupt handler that excelled only at talking to disk drives, namely DOS (see Chapter 3). TAPI was released before the advent of Windows 95 and Windows NT. Some serial communications and limited line device control is about as far as you will go on Windows 3.XX.

A powerful computer telephony system under the Win32 operating systems should reliably perform any of the needed computer telephony and switching functions concurrently, each process or thread of execution handled and managed separately, and on multiple telephone lines. In other words, it should be able to receive and make calls at the same time (on all ports). How do you program the device to do this? Just because the telephone and the PBX were designed for human usage, why can't someone just create software that makes a PC behave in the same way? In other words, let's get the PC to lift the handset (go off-hook), listen for dial-tone (perform CPA), dial some numbers, trap some numbers, and then wait for or generate (a) ringing or (b) busy signals (and again, perform CPA). When either of the tones are received, the software should be able to decide what course of action to take. This decision process is illustrated by Figure 2-3.

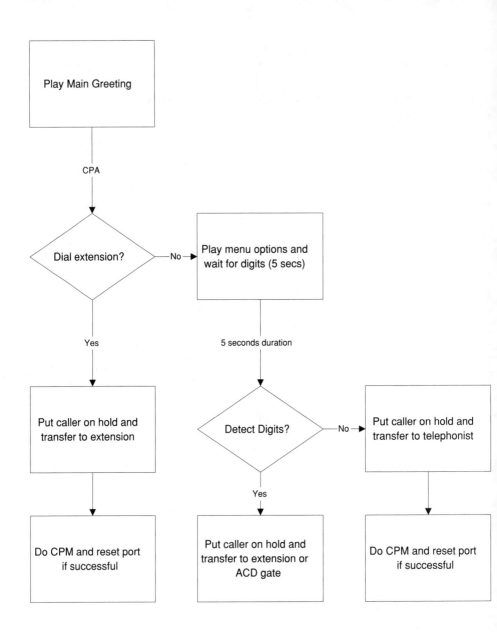

**Figure 2-3:** This is a simple automated attendant. Here, the CT port goes off-hook, listens for dial-tone (CPA), once dial-tone has been determined, dials the numbers. Then it attempts to detect ring or busy signals to determine the next action.

Sounds easy enough, right? Well, it's not exactly kids stuff. Some years back, just writing the software to perform the actions illustrated by Figure 2-3 could take and experienced software engineer, albeit new in telephony, several weeks to complete. (This was before the advent of computer-aided software engineering tools and toolkits that strip away some of the complexity of managing large software projects.) If the software had to work with a large number of PBXs, the development cycle for such a program exploded into several months, causing the poor engineer to work many late hours (which is why we have a bed in our office). Now we have TAPI, TSPAPI, the ECTF stuff and more to make the integration process a lot more organized and consistent (not simpler I must warn you). Do we just call the TAPI libraries for most of the telephony stuff and not worry about much else? Yes we do, and then some, as we show in Chapters 4 and 5.

But it doesn't end there. Once the gurus have released the software the systems integrators will have to attach the systems to their customer's PBX. If the PBX or CPE (customer premise equipment) is TAPI compliant the integration task may be no more difficult than connecting a printer to the network. But if there is no TAPI or TSAPI CTI-link the integration will have to rely on standard telephony signaling . . . in which case it may be a nightmare. As I mentioned earlier there are bazillions of different PBXs and telephone instruments of one type or another still in operation in the world at large. It is unlikely that these will ever be 100 percent TAPI compliant. They may need substantial modification to the proprietary software and electronics that control these devices. And so the task of integrating the modified setup into the existing information systems architecture and the LAN will remain a hellish nightmare for some time.

You, the forthright software engineer or, more than likely, the project coordinator, need to consider the viability of writing non-TAPI code as well. In other words, you may have to create a system that can achieve what it needs to with the native drivers provided by the hardware manufacturers, as well writing support to the TAPI libraries.

Granted, the awareness of computer telephony has risen at an astounding pace, thanks to the involvement of industry-leading computer and telephony companies and institutions, such as Microsoft, Novell, AT&T, Intel, IBM, ECTF, Rolm and others too numerous to mention. But these honorable enterprises are unlikely to recall a bazillion TAPI-less switches and offer every customer in the world a unconditional free "trade-up." I know one PBX-maker who has $30-million sitting in a bank account which gets larger every day because he has thousands of customers who are still paying off long leases running to the year 2010. My eyeball will become TAPI compliant before these PBXs do.

## Computer Telephony

All you need to achieve CTI is one small PC with a voice card or dialer and a small switch. But computer telephony integration means taking the PC (and all the telephony stuff inside) that you have integrated with the switch and attaching it to the LAN (sorry, intranet). Now this may mean your task is nothing more than a trivial addition of a PC to a network served by a single server. But it could also mean a not-so-simple integration into a highly complex collection of internetworking LANs and telecommunications services, consisting of many different types of servers, network operating systems, protocols, topology, and so forth. We explore this subject a little further in the forthcoming chapters.

A systems integrator or network engineer may have to perform two distinct tasks in using any product you create:

He or she has to get the CT system and the client's PBX and switch on speaking terms (that's CTI) and he or she has to integrate the CT system into the existing IT architecture without turning the enterprise upside down. That IT infrastructure may consist of a variety of client and server systems and mainframes, all interoperating in some way (well at least trying to).

I am always amazed at how the general Computer Press oversimplify the process of integrating computers and telephones, or invent things that don't exist (present company and publishers excluded). For starters, integrating the PCs, servers, LANs, and more with the telephone network (private and public) requires a little more than cursory understanding of both data and telephone networks and the enabling technologies (I hope this little introduction helps). As a result of a lack of understanding of the telephony side, very few IT executives have commissioned extensive computer telephony integration until their network engineers better understand the implications. LANs are hard enough to manage and maintain as it is, so few want to mess with what's working right now. Also the Internet/intranet technology is another factor to consider (see the introduction to TCP/IP and Winsock programming in the final chapter.)

Take the Internet for example: We have all fallen in love with the Internet; management loves it; employees love it. But the IT guru says "Wait a minute I am the sucker who has to integrate all this and make it work." Well, I have got news for software engineers and IT specialists: Management thinks getting voice mail over the Internet is a good idea too. And then there are a few cynics, or maybe realists, who still have doubts about the reliability of CTI. We will expand a little more on this in the final chapter, so hold that thought until then.

## First Generation CTI

You get CTI and then you get CTI. The wonderful voice processing card has been around for some time now. The principle means of "coupling" a voice processing or voice mail system, or any computer telephony system, to established telephony equipment in an enterprise has been via the analog channel on the mini loop that connects the telephone to the switch in the office. We call this practice in-band integration and, as I mentioned a few times earlier, you will be stumbling into it in this book if you plan to venture beyond the world of modem stroking and into hard-core computer telephony applications.

### In-band Integration

What do we mean by in-band? Take a look at Figure 2-4. It illustrates the bandwidth in the analog channel that spans the 300 to 3100 Hz frequency range. This is the range that is sufficient to transmit voice communications and sound. Now most of the installed telephony equipment in the world uses this same frequency range for control signaling. Control signaling is the process of using multifrequency tones to signal to the PBX or switch's CPU equipment the requests for dial tone, disconnect, conference, station busy, and so on.

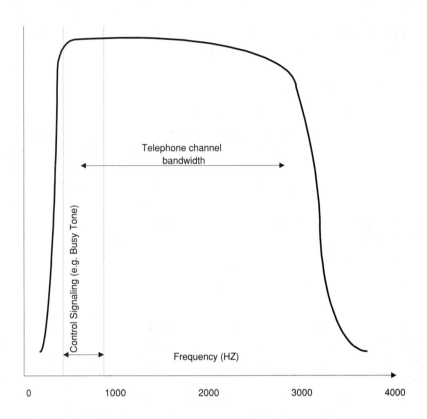

**Figure 2-4:** The voice bandwidth of an analog channel is capable of carrying conversation and control signals.

Now when you pick up the telephone handset and dial some numbers to transfer a call or put a call on hold, you're using this (audio) bandwidth of the analog channel to signal the PBX.

## Out-of-band Integration

In a digital environent, your telephone system delivers the control signals on channels outside of the voice channels. We refer to this in telephony jargon as out-of-band signaling. Using the out-of-band option for computer telephony is known as out-of-band integration. The digital (data/RS-232) connection between the switch and the computer telephony system also provides the medium for "out-of-band integration," although this is not the same as signaling "outside of the voice bandwidth" in a digital channel. Some modern systems, especially ACD systems, now provide Ethernet ports on the equipment for easy integration with a LAN or external computer-controlled data receiver.

Many PBX systems are quasi-digital devices that use digital electronics for switching and signaling tasks, but they use analog transmission electronics for transmitting voice. As alluded to earlier it is extremely unlikely that a conversation, especially a long distance one, will be all analog. The conversations will likely be switched back and forth from the two transmission technologies numerous times. Can you imagine if humans were cognizant of such switching of transmission technology. It would be like being aware of the process of seeing while trying to see.

Digital devices typically have a collection of several wires connecting the telephone sets to the switch. Almost all switch systems provide an operator's (I prefer the term telephonist) console to send and receive control signals to and from the PBX CPU. Again this is an external data connection, such as an RS-232 interface.

Although it is possible to communicate with the PBX by dropping and loading loop current, in analog environments almost all control communication is done by sending signals along the voice channel. Analog computer telephony systems thus integrate with the PBX and telephone systems by also sending and interfacing with the in-band signaling. Almost all call-progress monitoring done by computer telephony systems is done in-band.

No good computer telephony device monitors in-band analog signaling by intruding in a conversation. You simply plug the line into the port on the telephony card or voice processing board which deploys a signal processor and sundry other electronics to smoothly handle the signaling, telephony and processing requirements. In the U.S., these ports on the telephony devices, and all telephones, are fondly referred to as 2500 ports. This is Ma Bell's code name for a DTMF (touch tone) analog device. The old rotary- or pulse-dial phones were called 500 port devices.

## Call-Progress Analysis and Call-Progress Monitoring

Call-progress analysis (CPA) is the science of analyzing the signals on the telephone line in the process of setting up or tearing down calls. This capability should be nothing fancy for most CT telephony systems and telephony devices (such as modems) today. Your modem performs some CPA when it goes off hook and listens for dial tone. If it detects the (correct) dial tone, it dials the string of digits to set up the call and then listens for either busy or ringing tones coming from the network. A busy tone tells the modem to try later, and a ringing tone tells the modem to wait for answer or time-out after a period of ringing cycles (see Chapter 1).

If the remote device or network connection is lost, the modem, if it is so endowed, will again analyze the signal coming back from the network to decide if it should go back on hook. Some years back, the modems would remain on-hook long after a connection died and had to be powered down to put them back on-hook . . . an act of releasing a relay that bridges the circuit.

## Call-progress Monitoring

Call-progress monitoring (CPM) is the science of monitoring the progress of a telephone call between connection and abandonment. Again, this ability is not beyond the scope of functionality of the CT system. CPM, in other words, monitors and analyzes the signals that are active during the conversation between two humans and devices or both. Although the CPA and CPM may appear to be the same, they are not. Understand the differences before you sit down to design your computer telephony system. A CT system may use CPM, for example, to listen for a period of silence on the line. It then has to decide if a period of silence is the result of caller abandonment (when no other advisory signal is present) or if the people on the line fell asleep or died or are competing for the Olympic breath-holding trials. "In conversation" CPM is rare and in its simplest form involves three telephony ports or a three-way connection between the parties. The call progress monitor sits on the telephoney line and passively listens to the progress of the telephone call. TAPI, of course facilitates CPM without the need to have electronics "listening in" as you will see in Chapter 5.

Today, signal analysis is becoming so powerful that the device can detect and determine that although no one is talking, the background noise suggests that the parties have temporarily abandoned the handsets, perhaps to look for something, but did not go off-hook. CPM is most often used in human-computer dialog, to facilitate the recording of telephone conversations.

In many cases, the PBX or switch does not send any signals to aid in call-progress monitoring (of course I am talking about equipment that has not been TAPI blessed either). In these situations you will have to essentially guess what's happening. You can code the ability, depending on the functionality of the equipment, to detect human speech such as a "hello" on the other end of the line. This is often known as positive voice detection. Often, you can detect call status by monitoring changes in the loop current (again, if the equipment you are programming gives you this capability). When you encounter a PBX that doesn't have a loop current signal and that offers no in-band control signaling during a call, computer telephony integration is almost impossible. Or, at the least, the integration will be featureless. And if the systems do not interface (via TAPI) to the Windows systems services your life, as a computer telephony engineer, will be very difficult indeed. You may choose to just not support that equipment, or you may decide to knock up an electronic software controlled device that can be "bolted on" somewhere.

CPA and CPM are made possible by technology such as digital-signal processing. Without CPA and CPM, many of the most common computer telephony services, such as voice messaging and fax-on-demand, would not be possible. Humans have very advanced powers of deduction. We have been using the telephone, as a species, for more than a century now, and I believe that this telephonic prowess has been passed along from generation to generation in our genes, as part of our evolutionary process. I think Darwin called it natural selection.

You may laugh when I say that by the time you have cracked your third or fourth version of an extensive computer telephony system you can get on the phone and whistle to the switch for services. I know of one software engineer that can dial a string of digits with his voice (ok that's a slight exaggeration). We can tell the difference between background noise (like dishes being washed) and line noises on an abandoned call quite easily. But raw computer telephony components do not come blessed with this power of deduction. For starters they have to have the components to perform such analyses and you have to program in the logic to enable the system to interpret the information and act on it accordingly. You will almost certainly need a database of samples to provide the CT system with a memory cell of tones and signals to match samples against.

When a computer telephony system tries to place a call, the digital signal processor and supporting electronics on the computer telephony or voice processing equipment (usually the voice processing board) performs signal analysis. To do the analysis, the devices use in-band signaling. The results of the computations are then passed to other software control routines, which then compare the characteristics of the call, such as cadence and frequency, with signal information in a tone- or signal database (knowledge or experience).

The value ascribed to the result enables the CT software to switch to certain functions in the software, such as playing a file or starting a recording process. How does a computer telephony system know that a phone is ringing and that it should abandon after, say, three rings? As the decision tree in Figure 2-5 illustrates, when the CT system receives a tone, it compares the tone to the information in the tone table or database. The system can rapidly determine whether the tone is a ringing signal, busy signal, reorder tone, and so on. Then it checks the computer telephony application database to figure out what to do if it gets a ringing signal. The database may, for example, suggest the process listen for three rings and then transfer the call to the next extension, whose number lies in another memory cell (or a database field) somewhere.

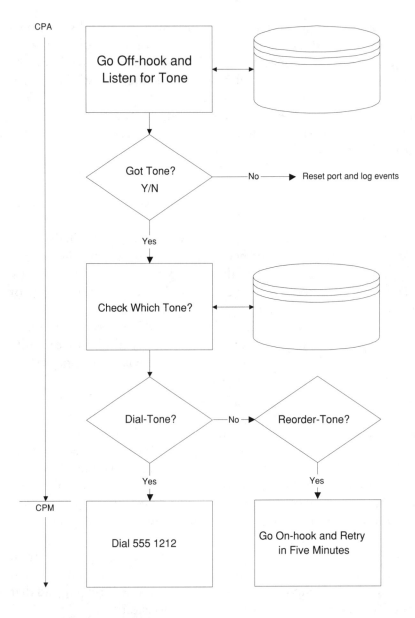

**Figure 2-5:** CT systems compare the tones they receive to program or data-based information to determine how to process the call.

In-band signaling (integration) is the chief means by which CT systems enable the caller to leave a voice message in the correct mailbox. Without an open partnership between PBX manufacturers and set standards to control how PBXs should work, the only way to do CTI with many switches has been by analyzing and acting on the divergent control signals.

Even when a PBX company has been hostile to the integration of computer telephony with their product, it still has been possible to do CTI. Some form of CTI is possible if the PBX uses the voice channel for control signaling and every PBX I know of does, even the "100 per cent" digital systems.

Modern PBX architecture enables the mixing of digital and analog phone cards on the PBX data bus. I have worked with a few companies that installed massive iSDX systems catering to more than 1,000 extensions. The iSDX is an ISDN switch. You can connect both ISDN and analog phones to this switch. At $500 to $1,000 a pop for an ISDN or digital phone, few companies want to hang a digital telephone off every extension. Without the voice channel and the control signal, the users would not be able to use the switch features. They would not be able to communicate with the switch, let alone make calls. I had two clients that use the iSDX. They have several thousand users hanging off their switches, all using plain old analog telephones.

So, the rule for integration (in whatever form it takes) is that if you can find an analog extension somewhere, some form of CTI is possible. You also have to remember that digital or ISDN PBXs and telephones do not operate in a world devoid of analog telecommunications. If an ISDN PBX requires a voice mail system, it should enable a caller on the PSTN to use an analog telephone to receive messages, even from a rotary dial telephone.

As far as CTI goes, anything that a user can do, the CT system doing call processing can do (often better); it just automates the process and clones the processing on every port available to it. The following process describes an automated attendant in action: When the caller connects and hears the greeting, he or she enters the extension number which the CT system, using signal recognition (DTMF/MF detection), matches against a database of extension numbers and users. The CT system then puts the caller on hold and calls the desired extension. While it's calling the extension, the system listens for the call control or call progress (in-band) signals, thus performing CPA. A short, fast tone means that the extension or station is busy or that the service is not available. A ring-back signal tells the CT system that the station is free but not being answered (remember Chapter 1?). If the system detects a busy signal, it will check its database to decide what to do next.

The next step may be to try another extension or to transfer the caller to a voice mailbox. Allowing the extension to ring continuously may provide the same result. The path that the call takes now is determined by the system administrator and the user. The information is kept in database tables for fast lookup. If the call is answered, the system goes off-hook, which connects the caller to the extension.

No two PBXs are the same. With all the hundreds of different models available from the dozens of manufacturers around the world, the integration process is a nightmare. It will probably take about two decades before all the closed switches (with no TAPI or TSAPI link) are finally put out of commission.

The following list is just a few of the control signal requirements that need to be taken into account when doing any in-band integration programming.

- The amount of time before loop current off is considered to be caller abandon
- Flash duration
- Tones to dial before a transfer is made
- Tones to dial after a transfer is made, before replacing the handset
- Tones to dial after a transfer is made, but the call must be retrieved
- Tones to dial before a blind transfer is made
- Tones to dial to recall a failed transfer to a busy extension
- Tones to dial before setting up a conference

But in-band signaling for CTI does not end here. Because some calls are diverted or forwarded to a CT system for coverage, the system first has to obtain prior information about the inbound call ringing at one of its ports, so that when it answers, the system can present the caller with the appropriate services. Why is this important? If a CT system does not have this information, it will simply go off-hook and default to the company or service's main greeting. This may not be a problem for, say, an audiotext, voice response or ACD service, but it will be a problem (a big one) for a voice mail service.

Without such information available to the CT system, the enterprise will have to rely on the human telephonist to connect a calling party to a specific mailbox just to leave voice mail. In other words, when coverage brings the call back to a live attendant, the attendant has to put the caller on hold, call the CT system, manually enter the correct mailbox number, and then replace the receiver. Only then would the caller hear the right mailbox greeting.

Of course, this is not practical and you do not want to force the caller to re-enter extension numbers at the message desk, especially if you are at all concerned about customer satisfaction. What a waste of human resources to have to do this manually too! This is technology gone bananas

So, how do you get this information to the CT system? One way is through in-band signaling. In this scenario, you have the PBX forward a string of tones to the CT system port as the CT system goes off-hook. It does this automatically by using a method known as repeat dialing (or digit echo, as it is known in Europe and English-speaking countries.) This enables the CT system to form an association with the call in order to identify it and link the call to a mailbox or a service. The process is as follows: When the CT system detects the ringing signal at the port, it goes off hook and listens for signals before playing a greeting. The PBX then transmits (repeat dials) the tones that represent the extension number dialed by the caller to the voice port. Using this method, the CT system knows which mailbox to drop the caller into and which personal greeting to play.

Repeat dialing is often used to enable users to depress a single, programmable key that calls the voice mail system, repeat-dials the extension number, and even transmits the passcode. Are you beginning to see how complex CTI can be? TAPI now makes it easier as we will see in Chapter 4 and 5. What if the PBX cannot perform repeat dialing, and a CT system does not have access to real-time station status information (digital) provided by the PBX? Then the only way to provide a voice mail service is to use voice mail in conjunction with what I call a computer telephonist (try translating the term automated attendant into Hebrew or Arabic).

A computer telephonist can perform this process because it handled the transferring process from the start, from the time the caller first connected to the enterprise, and attached a handle to the call. After the caller enters the digits corresponding to the extension required, he or she is put on hold and the system calls the extension. The computer telephonist now performs CPA (call-progress analysis) to detect an answer or busy signal. At this stage, the system is able to perform CPA because it is attempting to transfer on-consultation. Naturally, you have to disengage coverage on the extension; otherwise, the CT system will be merrily performing CPA when the PBX suddenly decides to send the call bolting off down the hall to another extension.

All in all, however, in-band integration is not the best way to connect a computer telephony system to a PBX. Often, using in-band integration is like having two people communicate in a common language that is the native tongue of only one of the parties. The receiving party may not understand 100 percent of the conversation because of differences in inflections and other idiosyncrasies that are not part of his or her culture.

The CT systems integrator has to remember that the integration method being used, in-band, was not designed for computer telephony. Many frustrating problems can arise as a result. For example: What if the CT system takes a message and then attempts to deliver it to a user? If the PBX has been signaled to divert the call into voice messaging, the CT system essentially ends up calling itself. It will likely (since computers are only as smart as the people that program them) begin recording its own message, kinda like the proverbial snake that eats itself. Though in-band integration will always be needed, we also have to look for alternative methods of integration. The best integration ability we have today comes from the APIs like TAPI.

Perhaps one of the biggest drawbacks to in-band integration is due to the critical nature of computer telephony in these mad-house days of call centers and the dearth of messages. A great deal of time is wasted when a CT system has to put a caller on hold and "test" an extension to see if the party is available. If the extension is busy, then a few seconds have been wasted while the system gets the tone back, compares it to tone or signal information, and then transfers the caller into voice mail. Multiply a few seconds by thousands of messages a day, and the seconds turn into minutes, which turn into hours. Wouldn't it be great if the CT system were cognizant of the extension's status before or as soon as the call arrives? The caller then would go straight to voice mail. There would be no clicking noises on the line, no waiting for the switch to respond to signals injected on the voice port.

## Digital Integration

Digital integration implies integrating or interfacing computer telephony with the PBX at the PBX's available data receptacle, which usually is an RS-232 interface. Two data streams of information are available that the PBX can provide at this port: (1) Station Message Detail Recording (SMDR) and (2) Station Message Desk Interface (SMDI).

The SMDR feed provides data on every call made by a user to the outside world and the duration of the calls. It provides this information for call accounting purposes. The SMDI feed provides the computer telephony system with information needed to present the caller with the correct mailbox.

The objective of repeat dialing and SMDI integration is the same: to tell the CT system who the message is for. The only difference is that the former information is delivered in the voice bandwidth of the audio channel, whereas the latter is delivered via an external digital link.

Several PBX companies have assisted the computer telephony process by modifying their software and enabling the ports to send more than rudimentary data to the SMDI or SMDR port. This data enables the CT system to determine who is calling whom. It tells the CT system that the call ringing on its port (associated with a PBX extension) was diverted from extension X. The CT system can then determine which voice mail box to open. This sometimes works better than in-band integration. But I would hardly call it state of the art.

## Summary

This brief sojourn in the world of telephony and CTI is somewhat historical. But, if you want to sell your turnkey computer telephony system or Windows telephony system to a wide audience, or if your are working on an IT department project, you will have to program in support for both TAPI and provide for the old fashioned CTI methods of in-band and out-of-band signaling. While this book will concentrate on what the Windows operating systems can do for you, the following chapters provide pointers to coding for both the new era and the old era of computer telephony under Windows.

Even in the US, technology leader of the world, there are millions of telephony systems that do not have immediate access to the wonders the likes of Microsoft, Novell, Intel, IBM, AT&T and others too numerous to mention are spitting out of their nerd nooks every day. Put yourself in the IT manager's shoes. Will you spend $100,000 to replace a PBX (and a likely additional $5,000 to install a Windows NT server) serving 50 people just to accommodate the purchase of a $2,500 computer telephony system?

Where to from here? If you are new to the Windows operating systems, the Win32 API and WOSA, you should read the next chapter. If you are not a rooky Windows programmer you can skip the next chapter and proceed directly to Chapter 4, which introduces TAPI.

Chapter 3

# Windows Telephony and CT Engineering

This Chapter Covers

- Mission Critical Services and Safe Systems
- The Telephony Operating Systems: Windows NT and Windows 95
- NT Unleashed: Symmetric-Multiprocessing Architecture (SMP)
- NT Unleashed: C2 Certified Security
- NT Unleashed: Support for Multiple Platforms
- The Win32 API
- File Handling
- Networking and Client/Server Telephony
- Networking: File Copy
- Networking: Mailslots
- Networking: Named Pipes
- Networking: TCP/IP
- Programming TCP/IP
- Remote Procedure Calls (RPC)
- Preemptive Multitasking and Multithreading
- Services
- WOSA
- TAPI
- Building Computer Telephony Applications
- Compiler Wars: "Battle of the Visual Masters"

Welcome to the World of Windows Telephony. The next three chapters serve as an introduction to Windows Telephony, and go into some detail about the Windows operating systems, programming for Windows and the Windows Telephony API, TAPI. But before we get into the technical treatise, let's get philosophical.

## Mission Critical Services and Safe Systems

You saw in the previous chapter just how complex CTI is. PCs and telephony systems operate in different worlds. The one half is an open systems platform, while the other half operates in a closed environment. Even with all the interoperability promises the telephony people have made, the telephony world (outside of PC and enterprise computing architecture) will, in my opinion, never be truly open, like the PC world.

So to build the true open PC-ACD, PC-telephony, PC-PBX, or computer telephony system, be it an IVR system or a voice messaging system what if we were to look to the PC world for the solution? The desktop or workstation computer is certainly powerful enough. A well-designed "PowerBX," "RiscBX," or a "AlphaBX" under Windows NT could blow away (in functionality and features) the most sophisticated of modern switches . . . in my opinion.

There are two questions about computer telephony I get asked repeatedly. Instead of just repeating the questions I have decided to be a little creative, Alex Trabeck style:

Answer: A modern operating system architecture that is robust enough to host reliable communications and telephony services.

Question: What is the Win32 System Services?

Answer: A modern operating system architecture that can be trusted to never crash and always provide users with communication and telephony services at their desktop computers, telephones, fax machines, pagers and beepers, home telephones, offices saunas, and more.

Question: What is the IT dream?

There is no such animal as a crash proof operating system. Every operating system crashes. I crash my Windows 95 machine daily, probably three to four times, especially when trying out some horrendous programming exercise that turns by debugger into pasta. I was visiting a Mac-based programming shop during the writing of this book. On the wall the McNerds pinned this joke: "Why did the computers on the alien ships in ID4 crash? They were running Windows." They probably were; they needed to only adapt one of the foreign language APIs and Windows NT would have been sufficient for them. I emailed this joke to a few of my NT friends, some at Microsoft. The replies I got was that the aliens had not invoked NT's security layers, and probably installed their systems on the older FAT (the old DOS file system).

You will hear this argument a dozen times and you'll continue to hear it, especially from PBX or telephony industry players. "The one thing going for PBX systems and the PSTN or CO systems is that they don't crash. Computers crash all the time. Not a week goes by without the LAN collapsing. Not a week goes by without an important file being lost or machine that locks up and goes on the blink for no apparent reason. I will not trust my telephony needs to a PC or a geek." Heard this one before, haven't you?

Yes, the PBX "mean time between failure" is astonishing. Most public switched networks are designed to keep running for 50 years without ever going off line. Many have exceeded that, even during equipment replacements and upgrades. They even keep on running during earth quacks and hurricanes.

OK, let's be fair you say, they have had three quarters of a century to get their act together. You will probably be able to cite situations in the world where telephone networks crash daily. And how reliable were these switches and things 50 or more years ago. On the contrary the PC industry is new. The operating systems are only now beginning to enter an age of stability and robustness and security.

You may have a point there arguing on the side of the PCs and their operating systems. But you can't dismiss the reasons people argue for the closed "indestructible" PBX. This argument is (seemingly) valid. That PBXs and switches are reliable is a reasonable statement. That they don't crash or can't crash is a lie. I have stumbled into PBXs that lose hundreds of calls a month (the owners just don't know this). I have tested computer telephony equipment on PBXs that can't maintain consistent frequency ranges and cadences (so you can't hang an automated attendant off them).

I have also overloaded some PBXs with so many calls, at such a high rate (like a new call every half a second), that the system runs out of memory and restarts . . . dumping dozens of callers in the process. And while consulting for Microsoft at its Johannesburg, South Africa office in 1994, at one time the PBX was down many more times in the month than the network.

But to follow this line is missing the point. It's unreasonable. PCs crash for myriad reasons, least of all the reasons is poor software and junky hardware. But PCs mainly crash because we use them so much, and they are so damn useful. They crash because we stuff all manner of hardware and software onto our systems to perform thousands of computations for all kinds of complex problems. Not a day goes by without many of us loading some new application onto the hard-disk, hardly a day goes by without a few thousand bytes moving onto and off our systems, traversing the Internet and the enterprise information network.

The PBX, on the other hand, is not something that every nerd, geek, guru, and garage-based teenage hacker can tamper with. Many people I know can look inside a PC and tell you what's cooking. For all but very trained technicians, peering into the internals of a PBX is like trying to dissect the human brain. The more modern switches have nothing more than a slew of interface cards and hundreds of strands of cable (the extensions) punched into the switching matrix.

The PBX or CO switch is a machine, another computer, that has a singular purpose to its existence. It switches calls. The PBX sits in a closet, gets dumped in a basement. Nobody, except the telephone engineers and PBX technicians, ever gets to look at it. Most employees don't even know where their company's PBX is kept; most employees don't care.

To a large extent it is understandable why the PBX has not come out of the closet. The telephone is the life-blood of almost every businesses. Your clients cannot afford to miss even one day of down telephone systems because it could ruin their business.

Some years ago I was in charge of the computer telephony system at a large Microsoft reseller. My company was also tasked with the responsibility of maintaining the telephone system. One weekend we brought in the PBX people to add more cards to the PBX to cater for an additional load of users and voice messaging ports. We left the engineers from a reputable PBX company (one of the biggest in the universe) to do their thing. The following morning, Monday, we heard they had drunk themselves silly while doing the reticulation. The result was down PBX for the morning. My client knows of at least one $100,000 order that went to the opposition resellers. Goodness knows how much more was lost.

It takes a human disaster like this to bring the telephony system at a busy company to its knees. What might it take to bring the computer telephone system to its knees. One nerd setting a network interface card to 8-bit support while the rest of the network is on 16-bit support (I have nearly done this too). As soon as that "flawed" PC powers up the entire network comes crashing down. It would then take a whole lot of time to recover the mess and collateral damage. No wonder the PC has a lousy reputation for stability. It's too easy to tamper and tinker with it.

There is another reason that many larger PBX companies have not implemented computer based switching systems: A resistance to change. Said one executive officer of a leading US PBX company: "What do we do with the billion lines of Cobol code we have. Dump everything and start all over again."

## The Telephony Operating Systems: Windows NT and Windows 95

So you think putting critical telephony and computer telephony services into such an open environment as personal computers is suicidal? Well, can it be done? The answer is a firm yes. You can isolate mission critical computer telephony systems from the main stream of networking and IT services (its just a matter of sensible IT management and philosophy). But even an isolated PC (or workstation) can collapse. So you need redundancy, fault tolerance, disk-mirroring, database and data back-up and archiving, log protection and more. I will make reference to system security and stability often.

The practice just described is not new. Many companies are now deploying systems dedicated to single tasks or applications. For example: running a client/server database management system, like MS SQL, in a busy company, is a mission critical operation. Many IT executives are dedicating a whole machine and Windows NT to this application. All that runs on the machine other that the DBMS, are services and standard facilities. You would do the same for a machine dedicated as a mail server, or a machine dedicated as a file server or domain controller or an application server. You need to do the same for your telephony server. Many companies thus have banks of high-end Windows NT machines dedicated to single applications and mission critical tasks. In the old days we would see banks of mainframes in the company computer room. Today we see the banks filled with NT servers, UNIX servers, Netware servers and the occasional OS/2 server.

In the PC or workstation world there are only three operating systems that I feel have the level of functionality and stability, that can be managed to stay up for an acceptable amount of time: Windows NT, UNIX and OS/2. Of these three systems only one can truly claim, in my opinion, to be really stable enough for mission critical computer telephony support at all levels, to cater to the full offering of Windows Telephony: Windows NT Server.

The reliability of the operating system on which high-end computer telephony systems are deployed is paramount. Windows NT is that reliable operating system, especially for running high-end telephony applications, such as switching, IVR, and call-center solutions. As mentioned earlier, computer telephony is becoming increasingly important at the desktop. Computer telephony is now very much about clients accessing critical telephony services via the GUI rather than the TUI (telephone user interface).

By using Windows NT server on the back end, you get unmatched desktop integration because both the client and the server are part and parcel of the same operating system family. (The Win32 systems services is at the heart of both client and server. For mission critical call-control, this is essential.) With the PC emerging as the key communications client, seamless connectivity to the desktop PC is critical. (I need to reinforce here that my affection for the Windows 32-bit operating systems has to do with what lies "under the hood." The applications front is becoming less important as the software applications models change.)

Windows NT is a more than a suitable telephony server for Windows clients, and given its open networking architecture, can connect to a variety of heterogeneous clients. Support for Transmission Control Protocol/Internet Protocol (TCP/IP)—which is considered by many IT managers to be the essential ingredient in enterprise-wide connectivity and the intranet revolution—began with NT version 3.5. I explore this topic later in this chapter.

TCP/IP can also provide access to TAPI 2.0 services on NT Servers. With NT comes significant remote procedure call (RPC) support, which is the capability for a client to invoke processors and procedures that execute on the server. These features are essential for logically integrating computers and telephone systems over the network. The following list provides some of the key factors that make NT a good choice as a high-end computer telephony platform:

- Cruiser-class client/server and true symmetric-multiprocessing (SMP) architecture.
- Scalability
- Preemptive multitasking and multithreading
- Redundancy and disaster recovery
- Great administration features, including the capability to remotely administer the system
- Big and growing array of "building blocks," such as voice processin cards, databases (SQL servers), messaging systems (MAPI), and so on.

What about Windows 95? This software is far and away the world's most popular operating system, with a market penetration that has well nigh exceeded critical mass. Windows 95 is a desktop operating system, more suitable for home and small business use (Microsoft now says, as marketing objectives for NT Workstation change), but actually it's a sufficient client for many large networks. Yes, it has its place in the computer telephony world. But not for mission critical telephony (server) services, such as switching services, PBXs, ACD, and Voice Messaging, IVR and more. Why? Read on.

Ok, so then what does Windows NT have that most other operating systems do not? For starters if you have studied the Win32 API as I have, you will realize that contrary to popular notion that Windows 95 is a DOS kluge it is really Windows NT with some high-end functionality stripped out, and a new GUI. The difference in operating systems is that Microsoft has chopped out certain areas that a client does not require. Specifically, Windows 95 lacks the following:

- Symmetric Multi-processing
- The full-blown C2 certified security implementation
- Multi-platform support

Many server oriented processes were also stripped from Windows 95. The remote access service was kept, but limited to only one session.

## NT Unleashed: Symmetric-Multiprocessing Architecture (SMP)

Out of the box, Windows NT Server 3.51 and version 4.00 up can support up to four processors (CPUs). But Microsoft created this OS for high-end portability and customization. Theoretically, no ceiling exists on the power of Windows NT. With customization and the right hardware (if it is available), Windows NT could provide enough processing power to run a destroyer or, in keeping with the subject at hand, to handle the most demanding of switching and call-center applications. To boost the power you just need to add CPUs.

If you are a little lost in the subject of system services and SMP it may help to discuss the operating system's microkernel. Every operating system has a kernel. This kernel consists of the minimum set of functions that have to reside in memory. Many computer telephony systems employ a similar kernel, not to replace the OS kernel, but to reside in memory and to be ready to process callers and handle tasks on the respective ports at any time. NT merely executes another instance of its microkernel on any of the available processors in the machine. In a single CPU system, milliseconds pass while an application waits for the processor to complete a cycle. In an SMP system, another processor comes to service.

Even with only one CPU on a high-end Pentium, NT will outperform its BIOSed interrupt-handling siblings. On other hardware platforms (this OS is truly portable) such as the DEC Alpha XT or IBM's Power PC, computer telephony applications will break the 30-port threshold in a single box, taking single box deployment to the hundred port threshold and beyond. It is thus perfectly feasible to anticipate the advent of such systems in the near term.

Windows 95 on the other hand only supports a single CPU. And of course it is bound to Intel architecture and is not portable to powerhouse computers like DEC's Alpha XT and the Power PC from IBM (which was running NT before OS/2).

Windows NT is also a multi-threaded OS (see below under the Win32 API section), and can thus can run OS threads on the separate processors. NT manages its thread allocation intelligently and can manage the resource allocation process to take full advantage of all the processing power available to it. It would thus not be difficult to develop a telephony application, such as an ACD, and allocate just the sufficient processing power needed for all the features of the ACD.

## NT Unleashed: C2 Certified Security

C2 security is really a subject fit for an entire book. In fact entire volumes have been devoted to the subject and the specifications. There are many issues involved in having C2 Certification; and, of course Windows NT is not the only operating system that can obtain C2 certification. Various flavors of UNIX come C2 ready; Novell touts C2 in Netware as well.

But no mission critical computer telephony or Windows switching system can afford to ignore security. Out in the field almost all of my customers would hit me on the question of security. You cannot build a computer telephony system without a security department in the package. You should be designing in security and access control, and event logging from the ground up. There are several security issues which have to be considered. Windows 95 has some security but it is not sufficient for advanced telephony like switching, object protection, access to communications mechanisms and more.

Here's a list of security concerns users the installed base will have:

- Mailbox protection
- Access control to switching services
- File access
- Operating system service access

One monster that surfaces in client/server, desktop telephony applications like third-party call control and desktop messaging is security of data. Windows NT provides some very capable security features for your data's protection. NT's security system is so capable, in fact, that you could go war with NT and be confident that your information was protected.
Windows NT in fact can protect data in accordance with the U.S. Federal Government's C2 and B security-level specifications.

These are mechanisms that were put in place during the Bush administration for the defense needs of the United States, especially for the protection and security of computing environments. What does that mean for computer telephony? Lots. Toll fraud and hacking is every IT or telephony manager's worst nightmare. It's a very real threat. A PBX running in a PC and providing switching to a corporation is likely to become the target of every hacker on the planet.

Windows 3.x, 95, and DOS-based computer telephony systems cannot adequately protect the sensitive messages of users from unauthorized access. Nor can administrators adequately screen certain users from messing with the mission-critical, call-control features that rely on information stored in databases.

In Windows NT, the files residing in a directory important only to the computer telephony server application are inaccessible to outsiders and unauthorized users. Security can be as tight as needed. Some applications require these files to be accessible only to cleared members of staff (via access control lists or ACLs). To the rest of the enterprise and the outside, hacker-populated world, the files simply do not exist. Of course you need to implement the NT File System (NTFS), to take full advantage of file security.

But NT security goes even further. You can even protect software objects, processes and threads of execution in Windows NT. Several years ago I headed a team of engineers that designed such security into a computer telephony system running on top of DOS. DOS itself has no security at all, so we had to build security to protect telephony services into the actual computer telephony kernel.

What we did was code the objects (in C++) to be password protected. The object itself, once accessed opened doors to various telephony services, such as switching. Take a trunk to trunk transfer. If a telephony crook manages to find out how to make a trunk to trunk transfer on your PBX he or she can essentially rape your company. Before you know it you have Iranians, Palestinians, Israelis, South Africans, Colombians, and probably a dozen other nationalities with private extensions and mailboxes on your PBX..

The objects we protected would typically give the user access to call transfer, audiotext files, voice mail boxes and many other possible routes and paths through the system.

## NT Unleashed: Support for Multiple Platforms

Another significant attribute of Windows NT, which counts in its favor as the OS of choice is cross-platform user-interface consistency. Sounds like a mouthful. What I mean is that no matter what machine, architecture, make or design of hardware you deploy, the user interface is the same. With the new Windows NT 4.0 shell released all the latest releases of Windows look the same . . . they all sport the Windows 95 look and feel, maintain the same system colors, software ergonomics, screen choreography and more (including bugs).

UNIX, on the contrary, can differ from machine to machine. There are also several flavors of UNIX (SCO UNIX is probably the most well known). Windows NT now runs on all Intel (high-end) computers, DEC Alpha XT platform, and the IBM/Apple Power PC systems. These high end machines have enormous processing power . . . they run at clock speeds of beyond 200Mhz and are getting faster all the time.

Having said all the above I guess you can summarize with the following analogies: Windows 95 is really NT Lightweight, Windows NT workstation is Windows NT Middleweight and Windows NT Server is, well, the Mohammed Ali of operating systems . . . "floats like a butterfly (on at least 32 Mbytes of RAM) and stings like a bee."

## The Win32 API

I was going to describe the services of Windows NT and along the way explain why it can be considered suitable as an architecture for mission critical telephony applications. Instead I will discuss several key regions of the Win32 API, what they brings to computer telephony, and point out the parts, with reference to telephony, that are not fully implemented in Windows 95.

When I refer to Win32 System Services, I am generally referring to core services that can be called via the Win32 API. When these functions or implementations are not supported by Windows NT client or Windows 95 I will make a point of noting this. For the most you can take it that I am not covering computing or programming against the 16-bit API and thus we are not discussing Windows 3.x or Windows for Workgroups.

Of course I cannot claim to be an expert of the Win32 API, and if you are serious about getting into some major Windows Telephony applications a full subscription to the Microsoft Developer Network is essential. It will be one of your cheapest outlays in your endeavor, believe me.

The Win32 API is massive, a popular adjective is "gargantuan." It contains thousands and thousands of functions that allow you to program against the system services and to create applications as well. Even when you use a visual development environment like Visual Basic, or a powerful object oriented language like Visual C++ or Object Pascal (Delphi) you are coding against the Win32 for calls to systems services, albeit you are abstracted above the API via "wrappers," interface libraries and interface classes. Most tools allow you to access the API directly.

The MDN includes volumes and volumes of documentation, so I will do no more than to introduce the API here and provide pointers for telephony applications. A good tip however is to work with the Win32 API help file that ships with the SDK. I find it easier to work with the online documentation because you can specify keywords to facilitate searching for the information you need.

Where do the telephony components fit into to the Win32 API? To understand how it comes together you should understand that the API is really composed of two distinct parts. At the heart of the API is the core OS services. Surrounding system services lie the supporting libraries. The supporting libraries have in the past been touted by Microsoft as separate toolkits and SDKs, and as components of the Windows Open Services Architecture (WOSA) we discuss a little further.

When TAPI was first released it was popularly known as the Telephony SDK or the TAPI SDK. When Microsoft now talks about the Win32 API they are typically referring to the core services and all the supporting WOSA API sets. Of course Microsoft may change things at any given time.

The following is rudimentary discussion of the core system services:

## File Handling

File handling is the essential and most basic service provided by all operating systems. Remember, that was what DOS was all about; a Disk Operating System to create and save files. File handling can be achieved at various levels throughout the API. Simple file handling, like opening a file, reading it, writing to it and closing it can be achieved with simple high level code.

But you will have to get close to the API when working with advance file handling, such as assigning powerful security parameters to a file, allowing say only a qualified process to open and read it, sharing files, locking, locking parts of a file (important in record handling, as you will later see). You will need to access the API to work with files containing binary data, such as object files that control the actions of your computer telephony system and voice files. The API has extensive support for file compression/decompression as well.

Another important region of the API has to do with file mapping. A number of functions exist to allow you to load a file into the memory allocated to a process. You can then create pointers and structures to access the information.

## Networking and Client/Server Telephony

The Win32 operating systems are designed from the ground up to turn the computers on which they run into communications centers. A computer telephony system running on Windows NT or Windows 95 can never be regarded as a "stand alone" computer or service. If it is connected to the telephone network it is a networked computer, and if it is connected to the LAN or intranet it is networked.

Note: Any machine connected to a network violates certain parameters of the C2 Security Specification, and thus may lose the certification. While Windows NT may be C2 OK out of the box, just connecting to a network is a security risk. See the C2 sidebar for more information.

The networking layers of your computer telephony system will be one of the biggest. You may choose or need to only provide limited LAN support and create, for example, a voice messaging system or automated attendant that only has telephone lines coming into it. But it doubtful that any computer telephony system today will be viable without LAN support, and the provision of "client/server telephony" services.

Remember what we talked about at the beginning of this chapter regarding

the stability of a computer telephony or computer based PBX. On the one hand you may feel it necessary to isolate the computer telephony systems from access via the network. But you would be depriving your enterprise of a wealth of computer telephony functionality at the
"desktop" by doing so.

For starters, let's say that you are building a voice messaging system. You will then be looking to integrate it to a PBX, hopefully one that is TAPI compliant. A switch that supports TAPI will need to communicate to your computer telephony system via the LAN; that is, via an Ethernet port, or a more direct CTI link. You could set up the computer telephony system and the switch on a dedicated LAN just to facilitate the sanctuary you may desire to create for mission critical telephony services.

However, be it a computer PBX or a voice messaging system one of the essential network services you are going to have to provide for is the notification of server generated events to client machines. What might these events be? Here's the two most popular uses:

PBX or computer PBX SMDI and SMDR information (see Lexicon): Your PBX will need to tell a separate process (on the local machine or on another computer) which extension was called (which telephone is ringing) or is being called. This service is essential to associate ring-no-answer (RNA) or station busy with the correct mailbox subscriber information.

As you read in the last chapter, the traditional means of handling this situation was through digital integration at, for example, the SMDI ports or via in-band signaling, actual testing of the telephone station. TAPI provides the interface to convey this data from PBX system to computer telephony system. You will, however, need to code in the appropriate network support to transmit the data from the server to the client, and have the server obtain acknowledgment from the client if necessary.

Another popular computer telephony service is known as the "screen pop." This is how it is supposed to work. A call arrives at the PBX or ACD and it is then transferred to an agent in a particular group. Just before or at the same time that the agent takes the call a window pops up onto the screen in which the agent obtains information or can insert information about the caller. Imagine that as soon as the call is answered the agent is able to greet the caller by first and last name and have on hand any information necessary for the caller.

Providing a screen pop and any other communication between client and server will require three separate processing solutions. First the information will have to trapped and associated with the caller (this association could happen on the server side or on the client side depending on the application, stress on the server, the objectives of the application and so on. This information trapping is not a difficult exercise because there is sufficient hardware and API material supporting the hardware to access Caller ID (Calling Line ID as it is known in Europe).

Then a data transmission facility needs to be opened to allow the server to notify the client, data or information is transmitted to the client. Finally the client needs to have the necessary functionality to stand-by on the receiving end of the communications process, receive the data, acknowledge reception if necessary, and act on the information (such as opening up a database record and so on).

High end computer telephony systems all live on database systems where telephone and computer telephony subscriber information is stored. Some of the many items of information you will need for regular access will be discussed a little later. A solid understanding of networking is essential for just about anything you do in Windows these days.

The paradigm of stand-alone computing is history. Sun Microsystems, Inc. made famous the slogan "the network is the computer." For computer telephony we might add words to have it read "the network is the computer telephony system." Figure 3-1 illustrates just such a computer telephony network. A solid understanding of the networking support inherent in the Win32 API is essential if you need to get into some serious low level coding of the various services. The following text serves as a very basic introduction to these services.

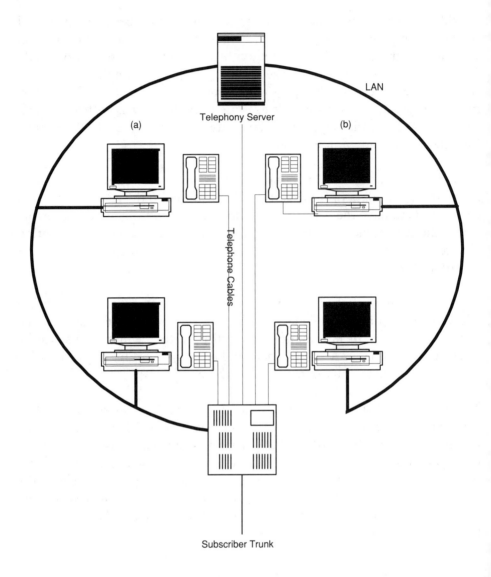

**Figure 3-1:** The network is the computer telephony system. This diagram shows a PC (a) connected to the LAN. It has a "logical connection" with the PBX (via the telephony Server). The client (b) has both a logical connection with the PBX and a direct connection to the telephone extension. The telephony server and the PBX are connected via an Ethernet interface. The telephony server may be the only server that such a small network needs—containing all applications and network serving software. Or it may connect to other servers, even a larger network.

## Networking: File Copy

File copy is the simplest function available to transmit data across a network. The advantage of file copy (calling the CopyFile function) is that it is very simple to implement in your code. The downside to using this route is that it is slow and cumbersome; really only suitable for the saving of large files of data. You would not want to send process control and procedure instructions data to a computer telephony client machine using this mechanism for several reasons.

For starters your CT client has no way of knowing when a file has arrived or has been updated without having to regularly poll a structure on the remote machine. Several years ago I led a team of engineers in designing such services to work on computer telephony systems running on DOS. While UNIX and OS/2 were certainly options for named pipes and other networking notification schemes the computer telephony manufacturers really only supported DOS so we had very little option but poll. What a pain. I longed for the advent of Windows NT and Chicago.

Polling not only eats CPU cycles (which may seem insignificant . . . it's not) this is not a means of "real-time" network communication. Remember how Microsoft Mail (or MS-Mail) worked. When messages were sent to your mailbox it took anywhere from several seconds to several minutes to be notified across the network. So a computer telephony system may need to call CopyFile to write voice messages, faxed files or an archive or back-up file to another machine, typically a hard-disk reserved for large binary format messages, but it is not a mechanism suitable for mission critical, robust, networking, supporting orthogonal communications.

Note: It may appear to be sufficient to copy and store binary message files in open directories, managed by an insecure file system like FAT. However, your security system is left "wide open" under such a schema. Many email systems are like that today, especially the Internet mail transport protocols (POP and SMTP).

## Networking: Mailslots

Another form of network messaging and notification is a mailslot. A mailslot is a unidirectional message transmission. It is generally used as a means of broadcasting messages to all machines at once as simply depicted in Figure 3-2. In order for the clients to receive the data they need to be "listening" to the mailslots.

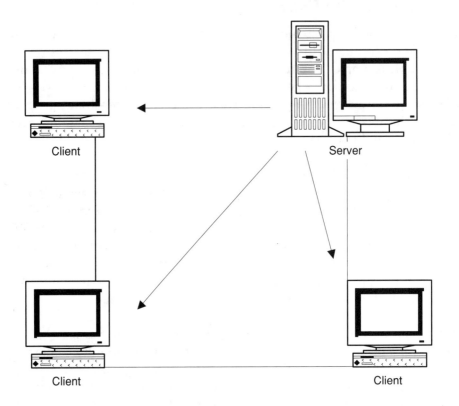

**Figure 3-2:** Mailslots (or UDP) in action. The server broadcasts messages to all machines (clients and servers).

Mailslots are useful if one machine needs to send a lot of messages to all other machines. The server of the message simply sends the message once and the entire network receives it. However, the server has no means of knowing if the message was in fact received on all machines. Of course the clients can act as servers and send their own messages to mailslots. But this is by no means a reliable method of intra-process (across the network information).

## Networking: Named Pipes

Named Pipes bring us closer the reality of real-time two-way or bi-directional communication between a server and a specific client on the network as simply depicted in Figure 3-3. A named pipe is in fact a mechanism to create a point-to-point connection between two applications or process on the network.

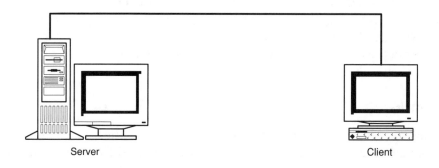

Server   Client

**Figure 3-3:** Named pipes (or TCP) in action. Server and client establish a point-to-point bi-directional communications mechanism.

Named pipes are useful in situations where computer telephony servers need to immediately notify individual clients about events that have occurred at the server. Such events could be calls that are in the process of being transferred to the telephone in or sitting next to the client workstations, message notification, and conferencing.

A good example of a named pipe application is a client that opens up a communication with a messaging server to invoke immediate message delivery . . . via telephone, fax, email or pager, and so on (provided, of course, if the client can be a named pipe server). The old way under DOS was to have the client "plant" a request in a message database (a duty roster file) that would be picked by the "sweep" of the message delivery service. The CT system would pick up the notification and then attempt to deliver the message.

Of course this was a typical DOS kluge, because under any network or telephony load the CT system would not be in a position to use any telephone or message ports to affect the delivery once it had picked up the notification. No, a client needs to "zap" the server via a point-to-point communication and instantly invoke a message delivery procedure. Named pipes are useful for such activity. In the UNIX world and over TCP/IP networks the TCP connection protocols provide a similar service to the Windows implementation of named pipe services.

It is important to note that named pipes and mailslots are functions of the Win32 API created for implementation on a homogenous Windows network. This means in essence that while you should use these services as you deem fit (and there will be times when you have no choice) they leave you stranded in a closed, albeit huge, networking environment. It will pay big to create computer telephony services that take advantage of the technologies emerging in the open system arena. This brings us to a discussion of Internet services such as TCP.

## Networking: TCP/IP

Millions of companies are connecting to the Internet. They use the Internet to exchange electronic mail with their employees, with their customers, with their clients, with their partners and to use the World Wide Web. And many more are now adopting Internet technology on their enterprise information networks and corporate LANs. The mechanisms of the WWW, particularly the publishing technologies like hypertext and the hypertext markup language (HTML) has led to a new model of information disbursement and access. These services transform the corporate networks into what are now affectionately known as intranets.

Microsoft has become extremely aggressive in its Internet/intranet strategy, which is welcome and exciting for the computer telephony geeks like you and me. (At first the Internet seemed to jolt Microsoft off its tracks, but contrary to many predictions, if there were any jolts they put the software juggernaut firmly on the Internet tracks. Microsoft probably now spends more money on the Internet than all other industry players combined.)

Why is the Internet, or, more correctly, the Internet protocols (TCP/IP) important to you. You see TCP as a network data packet transport protocol enjoys wide support in the industry. And the days of the self-contained homogenous Windows network are fast vanishing, which is not to say that this means that Windows is losing market share in any way. With the rapidly increasing popularity of intranets many different operating systems and machines will be communicating with each in myriad different ways. And the protocols of choice will be TCP/IP.

TCP is an able protocol for point-to-point communication across the network. Not only can you replace where necessary the named pipe service but the Win32 API contains extensive and thorough support for TCP/IP in its extension (WOSA) library known as the Windows Sockets API.

Computer telephony systems can thus communicate with not only many different clients but a wide range of host and server machines, message servers,

Web servers, PBX systems, database servers, main-frames and mini-computers, and more. Add the necessary routers and the computer telephony system can communicate with any machine in world connected to the Internet.

If you are a UNIX computer telephony programmer migrating to Windows NT you will be at home with the functions provided in the Win32 API because TCP/IP programming is TCP/IP programming no matter what the operating system.

As I mentioned earlier, I met many people at Computer Telephony Expo 1996 that were "stung" by customer requests for TCP/IP support. Many computer telephony software groups have postponed Winsock programming for years. Their time has run out.

Back in 1992, my company needed to build TCP/IP support into a computer telephony product to get the product to interoperate with mainframe computers running on DECnet via TCP/IP. To enable Windows applications to talk down to TCP/IP networks (from the presentation and application layers to the network layer), software needed to be hooked up to the Winsock network API. Needless to say, we had a lot of difficulty providing the service.

For the benefit of the newcomers to the Internet protocols here is a brief overview of the key components of TCI/IP:

The critical component of the Internet is its set of network protocols known as TCP/IP, which stands for Transmission Control Protocol/Internet Protocol. TCP/IP has been used in business networking on LANs and legacy networks for many years. Digital Equipment Corporation's (DEC) DECNET is a good example of an Ethernet LAN that has been running TCP/IP services for a while. Netware also had TCP/IP before any of my clients or I considered connecting to the Internet.

The TCP/IP set of protocols was initially developed by the U.S. Department of Defense to link different kinds of computers across an ocean of disparate networks, the result of DOD and armed forces tendering to companies, like DEC and IBM, sporting proprietary technologies. Once it was adopted by business, TCP/IP became the most popular suite of protocols to be deployed across Ethernet and X.25 networks. Although it has become the de facto suite of networking protocols, the successor to TCP/IP, which will be based on the Open Systems Interconnection (OSI) architecture, a child of the International Standards Organization (ISO)—is already being tested. For now, TCP/IP is suitable for the needs of client/server computer telephony environments.

(The work carried out by the DOD was at first successful because it provided for the delivery of the basic networking services, such as file transfer (FTP), electronic mail (SMTP), remote logon (TELNET) across a very large number of client and server systems. )

If the Internet is a train, then TCP/IP is its tracks. TCP/IP is the "New Master of the Domain" as LAN Magazine proclaimed in its October 1995 issue. As a result, corporations that have adopted TCP/IP as their chief LAN protocol (and many did years ago) have suddenly found a host of new products and development tools to install on their LANs that can make their people more productive. LAN managers and IT staff have, for example, found it far easier to set up a corporate Web site and give their users Web browsers than to maintain the old bulletin board systems that were so popular but hard to use in the late 80s and early 90s.

Also, BBS systems required dedicated connections from client to server over the PSTN, which is the chief reason why it did not become a pervasive technology (although popular), and why the BBS will be dead and buried by the end of 1997. Even voice-mail distribution lists cannot compete with the capability to publish information on a Web server.

What's more, much of the client Internet software is free, and the server software is very affordable. As a result, IT managers are bringing Internet technology into the business like bees bringing pollen into the hives. It's cheap, it's abundant, and it works. This is what intranetworking is all about. Intranets are nothing more than LANs running TCP/IP to support Internet technology such as Web servers, Web browsers and other Internet client applications such as FTP applets and data-rendering technology.

Office automation and groupware applications are being rapidly adapted to support intranetworking and the TCP/IP protocols. A whole new generation of software is emerging that has the Internet flowing through its heart and soul. I call it the "paradigms smashed" phenomenon of 1995, which had all the leading technology companies rethinking their strategic objectives. Now in 1996, many companies are doing things they never expected they would have to do. All software needs to be built on groupware principals and be communicative in nature, because the Internet enables even the loneliest computer user to connect to others and share resources. That said, managing a TCP/IP network is no simple matter.

Computer telephony is communications software, but without the Internet (and by default, intranet connectivity and support), computer telephony software is outmoded, outdated, and less effective. This point is where the new paradigms of Internet and intranet telephony come into being. I'd even go as far as to say that by 1997, any software company whose product cannot connect to the Internet and reliably talk to the new emerging Internet protocols will be dead meat.

Several computers in a small department can use TCP/IP along with other protocols, such as NetBEUI, on a single LAN. The IP component provides routing from the department to the enterprise information network or the WAN, then out to regional networks (WAN of WANs), and finally to the Internet.

During the Cold War years and earlier the TCP/IP architecture was designed to be robust enough to automatically recover from any node or phone line failure. This design allows the construction of very large networks with less central management.

TCP/IP, like all other communications protocols, is composed of layers:

The IP part is responsible for moving packets of data from node to node (computer to computer). IP forwards each packet based on a four byte destination address, which is known as the IP number or address. The Internet authorities, the Internic, assign ranges of numbers to different organizations and companies. And these entities assign the numbers to departments and workgroups. The IP operates on gateway machines, routers, that move data from machine to machine, department to department and organization to organization around the world.

TCP is responsible for verifying the correct delivery of data from client to server. But data can be lost in the intermediate network so TCP adds error checking to detect lost data and to trigger retransmission until the data is correctly and completely received.

So the Internet Protocol was developed to create a Network of Networks. What's an intranet then. First we connect our office machines to the LAN (via Ethernet or Token Ring). The LAN may consist of a Novell file server, a Windows NT server, Windows for Workgroup peers, and Windows 95 and Windows NT clients. Then we run TCP/IP as one of the network protocols on the LAN. A single machine can provide access to the Internet, using a router and firewall software that provide security to the intranet behind the walls of the enterprise. It is also possible to set up the network so that all machines within the organization are accessible from the Internet, but that would be a security hazard. It would be a good idea to set up a proxy server that allows client to access the Internet from behind secure lines.

## Addresses

Each technology has its own convention for transmitting messages between two machines within the same network. On a LAN, messages are sent between machines by supplying the six byte unique identifier. This address is known as the "MAC" or medium access control address, which identify the network cards. In an SNA network, every machine has Logical Units with their own network address. DECnet, AppleTalk, and Novell IPX all have a scheme for assigning numbers to each local network and to each workstation attached to the network.

On top of these local or vendor specific network addresses, TCP/IP assigns a unique number to every workstation in the world. This "IP number" is the four byte value that, by convention, is expressed by converting each byte into a decimal number (0 to 255) and separating the bytes with a period. For example, the UUNet server is 198.196.63.6.

To establish an Internet presence you would typically contact the Hostmaster@INTERNIC.NET or other network authority requesting assignment of a network number (the Internic is the administrative authority for the Internet in North America) or a domain name. It is still possible for almost anyone to get assignment of a number for a small "Class C" network in which the first three bytes identify the network and the last byte identifies the individual computer. You can apply to the Internic for any small network of computers in your organization. Larger organizations can get a "Class B" network where the first two bytes identify the network and the last two bytes identify each of up to 64 thousand individual workstations. Many ISPs get Class B networks and assign a range of numbers to their customers to use as subnetworks.

The organization then connects to the Internet through one of a dozen regional or specialized network suppliers, the ISP or IAP (access provider). The network vendor is given the subscriber network number and adds it to the routing configuration in its own machines and those of the other major network suppliers.

There is no magical mathematical formula that translates the numbers 198.196.63.6 into "UUnet Technologies." This information is stored in a network of distributed databases known as the DNS, or domain name system. The machines that manage large regional networks or the central Internet routers managed by the National Science Foundation can only locate these networks by looking each network number up in a DNS table. There are potentially thousands of Class B networks, and millions of Class C networks, but computer memory costs are low, so the tables are not unreasonable. Customers that connect to the Internet, even customers as large as IBM, do not need to maintain any information on other networks. They send all external data to the regional carrier to which they subscribe, and the regional carrier maintains the tables and does the appropriate routing.

## Subnets

Although the individual subscribers do not need to tabulate network numbers or provide explicit routing, it is convenient for most Class B networks to be internally managed as a much smaller and simpler version of the larger network organizations. It is common to subdivide the two bytes available for internal assignment into a one byte department number and a one byte workstation ID.

## Programming TCP/IP

In order to use the Winsock API there are a number of concepts you should understand. Once the concepts are fully understood you will find that TCP not only provides a viable alternative to using named pipes for computer telephony solutions but it is simple to write the necessary code in your applications. The beauty of TCP/IP is that you will be able to provide a computer telephony server with a world-wide knowledge base of every computer connected to the Internet, and the subnets and intranets.

### Concept One: The IP Address

As mentioned earlier every computer on a TCP/IP network is allocated a unique IP address and a domain name. The IP address is used by the server and clients to connect to each other in a bi-directional, reliable fashion. You will need to build support for this naming scheme into the computer telephony or Windows telephony servers and clients. They will need to access tables of IP and domain name information as needed to establish the TCP connections to each other.

### Concept Two: Sockets and Ports

Sockets, popularized by University of California at Berkeley UNIX, are a means of obtaining a handle to a bi-directional network service. Obtaining the handle and reading and writing to the service is almost identical to the obtaining a handle to a file that you can read and write to. The difference is that the socket you create is associated with the network and the transport mechanism, while the file is associated with the file system.

Once you have the socket you can either bind it to a port on the local machine, or via TCP you can connect it to a port on a remote machine. Once a port has been created and used in TCP mode it can "listen" for and accept connections. Any machine on your network can now connect to the port and send data to the machine hosting the port. Ports have queues that maintain the network connections. Your computer telephony client processes will establish a maximum size for the queue when they are enabled and start listening to the ports.

By writing this support into the computer telephony applications you have given the servers and the clients the ability to listen to each other and correspond. Servers can now send packets of data to individual client machines and the clients can respond in kind, with reciprocal data. This arrangement makes for the ability of the computer telephony system, such as a PBX or ACD computer telephony system, to send a client advance data concerning a call in progress.

We won't go into this in detail because how you do what you do with your computer telephony system, and the algorithms and solutions you create will be specific, and perhaps even unique to you situation and product. You can see, however, how (like the named pipe technology discussed earlier) a TCP connection using sockets and ports can trigger a screen pop on a agent's computer (the client). It is even possible for the client to respond to the communication as illustrated in Figure 3-4.

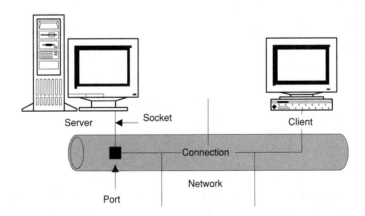

**Figure 3-4:** In this diagram both client and server connect to a port via the

socket that the server has created. The client can respond to the server via the same socket and port.

Now you can see how Windows has not only borrowed from the UNIX world, its adoption and built-in support for TCP/IP sockets gives it a robust alternative to mailslots and named pipes. In the TCP/IP or UNIX world mailslots and UDP (User Datagram Protocol) are alike in that they can be used to broadcast messages. UDP, like mailslots, is unreliable, but you can use UDP for a simple station/extension status application.

TCP, on the other hand, is a reliable point-to-point communication process that can be used, like named pipes, to establish continuing bi-directional and real-time communications between computer telephony
clients and servers as illustrated in Figure 3-5. Most UNIX programmers prefer to create TCP applications.

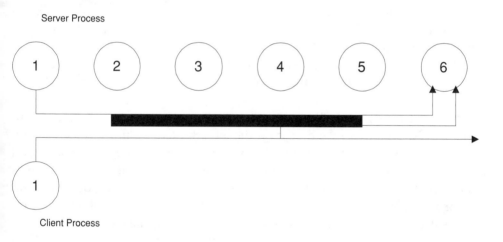

**Figure 3-5:** This illustration shows the server process creating a socket (1); it then binds it to a port (2) and starts "listening" on that port. The client process also creates a socket and connects to the port (4). Server and client can then begin transferring data between them (6).

Figure 3-5 illustrates how you can setup the client/server computer telephony system using sockets and TCP. First the server creates the socket and binds it to a port on the local machine. Then it enables it, puts it into listening mode and waits to accept a connection. As soon as a connection is detected data can be transferred between the server and the client. On the client side the local (client) machine also creates a socket and establishes the necessary support to enable the server to connect to it and transmit data.

To explore this idea fully let's turn our attention to the actual transmission process. TCP/IP networks can support three types of network packet structures. These are IP packets, TCP packets and UDP packets.

IP packets are commonly known as raw data packets that can be directed at a machine or broadcast on the network. They do not use the port arrangement described above and you would not want to use raw packets for most computer telephony setups.

UDP packets, again similar to mailslot packets, can be "shot" to ports on various machines or broadcast on the network. All ports listening will detect the packets and accept the communication.

The TCP packet mechanism, again similar to named pipe packets, are the data transmitted once point-to-point communications have been established between specific clients and servers. These packets can be aimed at specific machines on the network (to their ports) and these machines can respond in kind. Once you have mastered this communications technology I guess you can call yourself a UNIX programmer.

Let's zoom in on the TCP process. Figure 3-5 shows two machines, a computer PBX/computer telephonist on the left (the server) and a voice messaging system on the right (client). Both machines have enabled sockets and have established an on-going communications between them. The machines could be running software created by a single software vendor, or they could be products from different companies. (All three services could be built into one software package and deployed on the server.) They should still effectively communicate via TCP (which I consider even more important than first being named pipe enabled).

When a call arrives from the outside world it rings on a voice port on the server. The server harbors both switching cards and voice processing cards so it can provide both switching services and a computer telephonist/automated attendant services. A client enters the extension number of a representative and the server knows in an instant that the call cannot be transferred. The extension status indicates that the extension is busy and the PBX transfers the call instead to a voice messaging port on the client. (Remember from the previous chapter the old era CT systems would have to check the extension status by testing the extension and obtain in-band information or via an external data source.)

When the client receives the redirected call on its port it needs advance advice on the mailbox service to activate. It needs to know which mailbox greeting to play. By the time the port goes off hook the telephony server has already communicated this to client voice mail system. What took place was the following (refer to Figure 3-5): The server, configured as IP address 198.1.6.1 requested a connection to the voice messaging client (which is also a server) at IP address 198.1.9.1. The TCP sends a segment to the IP with information (flags) to indicate that this is the initial phase of the communication. IP receives the information and undertakes an address resolution process to determine the location (the location of the physical network interface electronics of the client). When IP gets this information it will transmit an IP datagram. At this point the server has hailed the client (in human lingo this would be the equivalent of "Yo, voice mail. You there?").

When the server (at 191.1.9.1) gets the datagram it transmits and acknowledgment segment back to the server effectively saying "Yo, server. That you." The server transmits a second segment to the client saying "Yo, voice mail, this is big S. Stand by." At this point in the communication a point-to-point communication has been established between server and client (stage 6 in Figure 3-5). Both machines are now ready to exchange data and more packets are transmitted. The communication will resemble the following: "Yo, voice mail. The call you just answered is for mailbox 414." (Of course there are several additional synchronization considerations.)

That's the process in a nutshell. Of course what we have discussed here is very basic indeed. Not only is the actual IP and TCP setup complex (there are several management issues) but the telephony environment will also bring problems and pressures to bear on the entire process. For example: Just switching the PBX hunt group setup from round-robin (clockwise) or sequential (to the pool of extensions dedicated to the voice messaging server) to random poses a different set of issues to resolve.

TCP/IP is an exhaustive subject. If TCP is new to you, you should get up to speed on the protocols as soon as possible. The future of CT is very much with TCP/IP. In the next edition of this book we will tackle CT-TCP/IP software in more detail. Perhaps the best book around that describes the Internet protocols is Richard Stevens' TCP/IP Illustrated, Volume 1, published by Addison Wesley.

Now that you have an idea of the advantage of having in place a point-to-point communication process between two machines we arrive at another critical service, relatively new in concept to CT, but enabled by TCP/IP (and named pipe) services . . . the remote procedure call.

TAPI NOTE: TAPI also provides methods for transferring a tele phone call and associated data to a client application (see Chapter 5).

## Remote Procedure Calls (RPC)

Ever wondered how on earth you can fire a procedure, trigger an event or process or call a function on a remote machine? Remote Procedure Calls (RPCs) do this for you. RPC allows you to distribute computing resources across the network. With RPC you can configure and set up services on machines to perform specific tasks. Not only can you distribute the processing load, but you can ensure that mission critical computer telephony functions on which the enterprise depends can execute on safe systems or in a safe environment (see the beginning of this chapter).

Consider the following necessary computer telephony operation. The PBX transfers a call to your telephone; you take the call but determine that it is meant for another party. Remember the basic telephony functions described in Chapter 1: You would have to flash the switch-hook, dial the new party's extension number, and either consult with the new party and attend to the transfer or transfer the party blind.

Now if you know the extension number of the party then the process is relatively simple. If not you will have to inconvenience your caller and transfer him or her to the operator or telephonist for assistance. Back at the "switchboard" there is still no guarantee that the caller will be speedily serviced. This situation is all to common in all enterprises. People spend longer on hold and call-blocking levels go through the roof. Corporations lose money when customers dangle in holding patterns unnecessarily. And in situations in which people move around a lot, or are never at any given extension for any length of time, such as in hospitals or hotels, the problems are compounded. Computer telephony can relieve this by employing databases, knowledgebases and CTI.

In the new era of computer telephony a PBX equipped with a CTI-link, TAPI compliant or otherwise, or a computer PBX, everyone in the enterprise can act as a live operator . . . by pulling up a GUI based PBX console. You could pull down address books and menu lists and even perform a transfer, a camp-on-hold, call park, or whatever.

One way to affect the transfer across the network is via an RPC, facilitated by a transport protocol such as named pipes or TCP. The process, as far an engineering is concerned, is relatively simple (at least to imagine). The end in mind is identical to the end result of flashing a switch-hook, signaling a PBX CPU for service, and affecting a transfer via in-band or out-of-band signaling (see Chapters 1 and 2). You are performing the tricks via the computer network rather than via the telephone network.

Instead of calling a function on the local machine you call the function on the telephony server or to the PBX directly, if it is running on Windows NT, or another suitable operating system. The procedure executes on the remote machine as if someone sitting at the remote console executed the command from the user interface of an attached console. Of course you will have to write software to explicitly bind to the remote machine to establish the link; but the route is pretty straightforward.

Many enterprises are considering RPC because it allows them to perform mission critical functions on reliable architecture. A typical RPC might be a safe, fault-tolerant multi-processor system (a combined throughput in CPU power above 250MHz), running Windows NT, and executing procedures in hundreds of Mbytes of RAM.

Note that Windows 95 does not support named pipes as a server. It can only behave as a named pipe client. You may have to turn to TCP and other mechanisms if your applications calls for Windows 95 clients to behave as the appropriate communications servers. While it all looks relatively easy setting up RPC services is complex and demanding. You also have to remember that the clients and RPC servers are now transmitting parameters up and down the network which will have an impact on the network because of the combined overhead. And don't forget you have be sure that the implementation will not impact the "safe" margins of operation you have set for enterprise.

## Preemptive Multitasking and Multithreading

If you are new to the Win32 API and the concept of multiple processes and threads then the following introduction will help focus on these issues.

Let's look at DOS telephony applications for a moment. Available device drivers, development tools and a programmer friendly development environment have been the chief reasons many telephony applications have been offered on DOS. Memory, in certain respects, was not really a limitation in recent years.

The clever engineers just sliced through that barrier with DOS extenders.

We put the device drivers into the base RAM and put the telephony application in extended DOS. Then what did we do? We created lean "telephony kernels" to sit atop the DOS kernel and essentially bypass the DOS chain of command. To speed up processes in computer telephony applications, programmers eschewed accepted protocols and bypassed the input/output system, accessing the hardware directly and weaving in and out of undocumented backdoors like mice navigating Swiss cheese (see Figure 3-6).

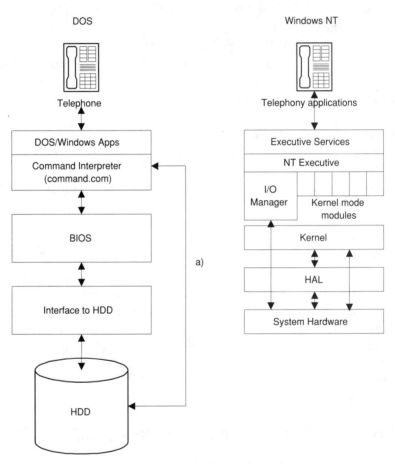

**Figure 3-6:** The old DOS architecture on the left allowed programmers to access devices directly (a). Under Windows NT applications need to make client/server process requests through several security layers to gain access to "kernel mode" processes and devices.

What you essentially have here is a breakdown in protocol and the shunning of OS policy and security. The crew takes over the ship, locks the XO in the bilges and takes the vessel into uncharted waters. That's when things start to go wrong. Take the submarine movie Crimson Tide: Without the resistance (the vested authority) of the XO, the rebellious submarine crew would have launched their nuclear missiles and erased our planet from the universe.

The DOS ship was not designed for mission critical applications, under severe stress it cracks up, especially at the file system level. Any IT executive worth his or her salary knows that when a VAR brings in a CT application under DOS, claiming to handle mission critical telephony, processing more that 12 or 16 ports concurrently in a single Intel class machine, some serious tinkering has taken place under the hood. Now who will guarantee system integrity? Who's prepared to guarantee that the system will not crash, under any circumstance short of ripping out the power cord? In my experience these applications in high traffic, high end environments are glass houses, waiting for single pebble (like a cross-linked file or a corrupt record lock file) to trash everything.

Computers waste a great deal of time. For enterprise-wide IT needs, they waste too much time. In DOS, when a request is made to read or write data to a hard disk, the CPU (which operates on to a nanosecond clock) goes to lunch while the hard disk (which operates on a millisecond clock) does its thing.

In Windows, a component known as the Task Scheduler (or task switcher) makes sure that the CPU gets no such breaks. The CPU always has work to be done. Perhaps an appointment needs to be scheduled, or some high-priority calculation or memory scan must be done.

A number of applications may be running in NT or Windows 95. The voice mail administrator may be adding users, or a remote engineer may be performing some maintenance (connected via the built-in, remote-access service or via the LAN or intranet connection). The Task Scheduler keeps the CPU working on background and foreground processors, ensuring that foreground processes get higher priority. So, when a caller dials in and wants to listen to a ten-minute voice recording, the CPU can be engaged with other tasks while waiting for the drive to access the data. After the data becomes available, the processing resource can be switched to the I/O needs and the requesting process will be serviced.

Computer telephony applications, whether voice mail or full-blown switching services, are I/O intensive. Disks are constantly on the move. Over the years, users have learned to squeeze as much juice as possible out of single-process computers. But the database and disk-thrashing activities in messaging and, particularly, computer telephony applications are rocketing in accordance with user demands for service.

Every time a caller makes a request to be switched, leaves a voice message, or checks a bank balance, the machine experiences a substantial I/O strain. With enough RAM, it is possible to load into memory all the necessary data (such as pointers to essential greetings or even the greetings themselves), but the application remains restricted to single-task processing.

Under Windows 3.X a form of multi-tasking was possible as Windows "allowed" several applications to execute cooperatively. This is known as non-preemptive multitasking. Each application that is given foreground priority essentially takes over the CPU while the others "sleep" in the background. There is no-ways that this can be considered a safe computing environment for mission critical application because should the foreground process crash and hang, the entire system hangs.

Under Windows 95 and Windows NT multitasking is preemptive. This means that when a process requires access to the CPU the operating system preempt the processing application and allocates "time" to the other process. The OS is preempting the executing process. This also means that the OS is in charge of the running of the house-hold. Unruly or corrupt processes can be terminated in the event they crash or threaten system integrity.

Threads add to the concept and power of multi-tasking under Win32 in that applications can divide themselves into threads which are then given CPU time as needed. The Win32 operating systems are themselves multi-threaded.

NT, in that the operating system itself uses multiple threads, has the capability to clone as many copies of its microkernel as needed to match the number of processors in the system. The combined throughput from high-end processors is astonishing. A database application that reads and writes data through the IVR system's interface can delegate essential services to threads that for an available processor to execute. Thus, a well-written NT telephony application can merely apply for more horse-power as it needs it, regardless of the amount. This capability, built into the very "soul" of NT, is not available in any other OS.

Let's look at this in the telephony environment. While the Task Scheduler is making sure that the CPU keeps busy, one caller waiting for a prompt or information will not notice any decrease in service. But sixteen callers accessing an IVR application running out of a single box are bound to notice that things are getting a little slow; and the pauses between prompts taking longer. A multithreaded database application sitting under a computer telephony interface performs database I/O for several client processes simultaneously.

Instead of waiting while a single thread completes a database read/write, another thread can be writing pointer records to voice messages stored in the message repository. If you now take all the capabilities available to a single operating system described earlier and put the OS (Windows NT) to work on a multiprocessor system, you end up with the horsepower and throughput for the highest of high-end computer telephony applications.

Both switching and human-computer dialog can be provided in a machine that is, essentially, a mainframe in PC clothing. Under SMP, database and I/O-intensive applications are catapulted to processing levels that far exceed anything possible with DOS, Netware 3.x, and even several flavors of UNIX. (Netware 4.1 has its similarities, including SMP; and of course, UNIX and OS/2 also are multitasking operating systems, but I won't delve into OS comparisons here.)

## Additional Failsafes: Redundancy and Fault Tolerance

The Windows operating systems also have other attributes that make them more than suitable for high-end, computer telephony applications. Not long ago, providing a client with redundant systems and sufficient disaster-recovery and fault-tolerant mechanisms was expensive and labor-intensive. Such items are absolutely essential in computer telephony systems, however. After all, you cannot afford to have the PC PBX go down and lose all the call-control records. The beauty of Windows NT, in this regard, is that it throws in the necessary software for these mechanisms as part of the deal.

The computer drives (hard-disks, floppy drives, tape drives and CD units) are the weakest components on the computer, because they have moving parts. The NT file system (NTFS) redundancy is provided through transaction logging. All transactions or operations on a hard disk are logged to a special log file, which ensures that the hard disk structure (not a dead disk) can be recovered. NT also achieves its redundancy and fault tolerance through mirrored disk partitions, RAID5 support for disk arrays (the computer industry's latest specification for establishing redundant arrays of inexpensive disks), and disk duplexing.

The NTFS is a fully recoverable file system. With this system, cutting the power in the most intensive database or disk I/O results in no more than the loss of data that was being processed. The system restores to its previous state. To prevent the loss of data integrity, you merely need to make sure you have activated NT's built-in redundancy options.

## Services

Before we switch to WOSA and TAPI a discussion of Windows NT Services will be invaluable in understanding how TAPI services interoperate, as described in the next chapter.

Services are the executables that need to start up at OS boot time and always remain in the background to service applications and processes. Examples of services are the Internet Information Server, the FTP Server, the Remote Access Server (RAS) and, in line with the subject at hand, the Windows Telephony Service.

The concept of services is not unique to Windows NT (although the full extent of the technology is not implemented in Windows 95, which loads services via the "Startup" folder). UNIX has an admiral reputation for handling services, which it does via processes known as Daemons. Both Daemons and Services are similar in that they load up at boot time and stick around to handle tasks. The most well known of UNIX Daemons is the mail Daemon, Sendmail. These faithful executables are what keep the Internet mail flowing around the world.

A PBX service in a Windows NT machine will be the controlling executable that manages the PBX equipment and cards that are either on the Windows NT bus or attached to it via cables or LAN connectors. The service, along with the OS Telephony Service, will always be running, no who or what is operating on the machine. A Windows NT machine quite admirably handles key telephony services while servicing a dozen users who are logged on via the RAS service. An RPC/telephony server, as discussed earlier, should be created as a service so that it is always running and thus always available.

Services can be stopped by an administrator without powering down the NT machine via the Server Manager dialogs. At any given time Windows NT itself is running services for essential OS processing, such as responding to messages and the clipboard server.

Creating services and synchronizing them with the operations of the operating system is not the same as creating applications that you run from the NT command line or user interface. Services need to be installed and placed under the central authority of the Service Control Manager. No, this is not some guy or gal sitting in a back-office at the sprawling Microsoft campus in Seattle; it is an executable in Windows NT that maintains a list of all services and their start-up status.

Naturally, service executables reap all the benefits associated with the new 32-bit technology offered in the Win32 API. In other words your service executables can, and should be powerful, multi-threaded applications doing all manner of telephony operations, such as voice processing and fax processing on multiple ports and lines.

## WOSA

Microsoft has invested considerably, and "banked" heavily, on the hope that business and enterprise would adopt the Windows operating systems as the "strategic" platform for client/server computing. In order to execute this mission however an open architecture needed to be created and adopted by service providers and application providers. Within the architecture Microsoft would provide a collection of APIs against which independent software developers or vendors (ISVs) could create and sell product. The underpinning architecture is known as WOSA, which stands for Windows Open Services Architecture.

WOSA is a constantly expanding and encompassing architecture. It is constantly being extended and reworked to meet changing conditions on the enterprise battle fields and on the technology front-line. The critical advent of the Internet is a good example of how the WOSA shifted and expanded to provide key APIs and SPIs, such as the Internet Information Server API and the Telephony API. Windows Sockets and the RPC mechanisms discussed earlier are key components of WOSA, as, of course, so is TAPI and MAPI, the Messaging API.

TAPI is not concerned with providing access to the information exchanged over a call, the so-called media, or information streams. Rather, the call control provided by the API is orthogonal to the information stream management. The Telephony API can work in conjunction with other Windows services, such as the Windows multimedia wave audio, MCI, or fax APIs, to provide access to the information on a call. This setup guarantees maximum interoperability with existing audio or fax applications, Internet telephony, distributed computing, intra-networking, and more.

## MAPI and other Important Technologies

Before we shift gear and launch into TAPI, I would like add that APIs such as MAPI and ODBC, and other key technologies, such as DAO (the Data Access Objects), OLE, and the Component Object Model (COM) are as essential as TAPI in creating robust, powerful and useful Windows Telephony products. Rather than a hastily include reference to these APIs, which would have delayed this version of the book, I held them back for future editions, volumes, or titles, which I hope to bring to you.

I consider the whole MAPI (simple and extended MAPI) to be an essential component of any Windows Telephony application. There has been much written about MAPI, especially simple MAPI, and of course about Microsoft Exchange, a messaging server and client environment. The MAPI is a huge library of functions, the largest component of the Win32 API. So big is it that many engineers run like crazy when asked to program against it. Many refer to it as the "API from Hell." There was much I wanted to accomplish, as the CEO of a software development company, with MAPI, but it was no easy task putting together An application programming team to do this. So instead I assembled the necessary resources to build higher-level interface tools that all application developers can use.

MAPI is not nearly as intimidating as it looks when you break it up into smaller logical components. For starters there are parts of MAPI, like schedules and forms that you may not need to address in your product. Be that as it may; I would like to add that my company has spent most of this year creating these high-level toolkits for MAPI that can make the task of the engineer a lot easier (with less cost and considerable time-saving). Towards the end of 1996 we'll have much documentation and techniques in hand that will enable you to build high-end messaging systems integrated with telephony systems for the likes of Visual C++ and Delphi; that is, if extending the Exchange server and client is not on your hot priorities list. If you have any questions regarding MAPI, or anything we have discussed in this book, do not hesitate to give me a yell at js@wizzkids.com . . . or drop by our Website at WWW.WIZZKIDS.COM.

## Building Computer Telephony Applications

While the next chapter delves into the Telephony API in considerable detail it is worth your while to consider adopting a visual telephony application development environment such as Visual Voice, a product created by the Stylus division of Artisoft, Inc. Stylus, formerly known as Stylus Innovation, Inc.

offers a collection of telephony and TAPI tools that can take you considerable distance in developing voice processing and computer telephony applications. The toolkit is essentially a collection of OLE controls (OCX components) that encapsulate the key calls to the telephony functions of TAPI and other API sets, the likes of which are published by Dialogic Corporation and IBM.

You, the developer, typically access this functionality via published properties, and handle applications specific routines via events triggers generated by the controls. Many of the properties can be accessed at design and run-time and the controls can also be cloned in separate threads for multi-line applications.

While Visual Voice was initially created for Visual Basic programmers, the implementation of the control as 32-bit OLE components gives the Visual Voice tools a whole new meaning in my opinion. You see, in keeping with my philosophy of safe programs and systems, I have never considered Visual Basic a programming language in which to create mission critical telephony applications. I have never regarded it as a true object oriented system, although the recent versions are nothing to snort at. My problems also had a lot to do with versioning (keeping track of the VB runtime libraries) because Visual Basic in an interpreted language.

I have over the past two years studied Borland Delphi and Object Pascal language extensively. From the launch of this product back between 1994 and 1995 I fell in love with the purity of the Object Model in Object Pascal. Object Pascal is also extremely easy to read and thus makes it easy to document extensive projects. Once hooked on Object Pascal and Delphi's visual programming environment you may turn your back on C++ for good. Although both have their strengths and weakness. It is worth while to maintain or develop a engineering corps that has strength in both languages. Object Pascal and Borland C/C++ also share the same "backend" compiler technology and you can quite gracefully mix and match your object code as you deem fit.

The good news about the Visual Voice components is that they can be accessed without much problems from Delphi (you have to hide the controls on the form though because they do not become invisible at runtime). This means that the Stylus tools are really now (Windows) programming language independent.

Rather than discuss the pros and cons of the compilers I have left this task to a popular article written by DBMS Magazine's David Linthicum. However, to demonstrate not only the power of Object Pascal and Delphi, but the Visual Voice components I have included the commented source code of a Delphi multi-threaded application that clones the VV components at runtime in a multi-line voice processing application as Appendix B.

## Compiler Wars
### Battle of the Visual Masters

By David S. Linthicum
DBMS

**Delphi vs. Visual Basic vs. Power Objects.**

In early 1995, as Microsoft's Visual Basic entered its fifth year, the crew in Redmond found its dominance of the visual client/server rapid application development tool marketplace challenged by Borland's innovative Delphi 1.0. With every capability of Visual Basic plus a few, Delphi exploded onto the market with such a bang that it sent the boys in Seattle back to their compilers. Thus, we now have the Windows 95-compatible 32-bit release of Visual Basic, release 4.0.

There was, however, another challenger waiting in the wings. Oracle's very own Visual Basic "wannabe," Oracle Power Objects 1.0. Power Objects is clearly Oracle's attempt to build a Visual Basic clone, and its hope is to lure existing Visual Basic developers into its waiting arms. Power Objects is the youngest of the three tools, now at version 1.0.

Not to be outdone by Microsoft or Oracle, Borland again came on strong with its 32-bit, OLE-enabled 2.0 release, the newest revision of the bunch. Microsoft is already working on Visual Basic 5.0, and it shouldn't be long before we see the next version of Oracle Power Objects as well. The battle of features and functions continues.

If it sounds like the type of battle you see between long distance companies, you're right. This is a battle for the hearts and minds of client/server developers, and the combatants are willing to do almost anything to win. The question you need to answer is: Which tool is right for your client/server project?

Just for the fun of it, let's place Visual Basic, Delphi, and Oracle Power Objects side by side. Let's test development capabilities, database support, performance, and other features that really matter. May the best tool win! (I should state here that my description of Power Objects in this article is based on a version I downloaded from the Web, because Oracle's public relations representatives failed to release a shrink-wrapped version of Power Objects to me in time for this article.)

**Battle Royal**

From 35,000 feet, these tools look very much alike. All provide advanced visual development environments, the ability to use OLE components (OCXs), built-in links to a variety of database engines, and an underlying programming language. Upon closer examination, however, the similarities abruptly end.

Visual Basic was one of the first rapid application development (RAD) tools that brought drag-and-drop component-based development to Windows 3.1, and now to Windows NT and Windows 95. Visual Basic's experience shows in the current release, which rides on OLE exclusively. Using the power of OLE automation, Visual Basic developers can not only access and link to OLE automation servers, they can create OLE automation servers (in-process and out-of-process) that are accessible from other OLE-enabled applications. What's more, Visual Basic developers can create automation servers that are accessible over a network, and thus provide an application partitioning mechanism for multi-tier development.

If you know how to use Visual Basic, you'll have no problem using Oracle Power Objects. Unlike the other players, Power Objects supports both Macintosh and Windows environments, but it does not offer support for the politically correct ODBC. It provides native database drivers only. Power Objects uses Visual Basic's application development language (Basic), and can use OCX components as well. There is no VBX support, however.

Delphi is the radical of the bunch, with Object Pascal as the underlying development language. Like Visual Basic, Delphi (with the 2.0 release) provides both in-process and out-of-process OLE automation services, and can use OCXs in its applications. The most important feature of Delphi is what it doesn't do; namely, it does not use a p-code interpreter. Both Visual Basic and Power Objects require that runtime applications pass through an interpreter, which means slower applications and more system resources. Delphi compiles its application to an optimized native executable that runs like the wind, and eats fewer memory chips and processor cycles.

All of these products come in different versions that enable developers to match the tools, and the price of the tools, to the application. Power Objects comes in both a client/server version (with links to database servers) and a standalone version. Delphi offers Delphi Desktop for individual developers,

Delphi Developer for professional LAN-based developers, and Delphi Client/Server for client/server developers. Visual Basic offers similar packages, including a Standard Edition, Professional Edition, and Enterprise Edition.

**Component-Based Development**

All three tools support component-based development, or the ability to build applications by assembling component parts. In fact, all of these tools rely more on component-based development than on object-oriented features such as inheritance and encapsulation. This seems to be an easy development model for beginners to follow. However, each tool supports objects in its own special way.

Component parts are either native components (or objects) that are not portable from one tool to another, or third-party components (such as VBXs or OCXs) that are portable from tool to tool. The general idea is to use as many prebuilt components/objects as possible before creating your own native components, or to use third-party components to add flare to your client/server application, such as including a calculator, VCR controls, or mainframe access.

Third-party components once dominated by Visual Basic Custom Controls (VBXs) gave way to OLE Custom Controls (OCXs). Why the change? VBXs were not based on a standard component-based development platform, and thus declared a "kludge" by those tasked with building these mini applications. OCXs, in contrast, use standard component-construction mechanisms and interfaces based on OLE. Visual Basic still supports VBXs, as does Delphi; Power Objects supports OCXs only.

Delphi's component-based development architecture provides developers with the best of the object-oriented and component-development worlds. Delphi is written using its own library of components and tools. Application construction is simply a matter of building upon Delphi's application framework. This framework, called the Visual Com Library (VCL), is a set of object types that provide the foundation for an application, so you don't have to build objects from scratch.

The VCL provides everything you need to create user interface objects for viewing and changing data, list boxes, combo boxes, and menus. The VCL also provides a basic set of components such as grid, tab, and notebook objects. As with Visual Basic and Power Objects, Delphi developers select components from a palette and drag them onto the application work area.

Delphi lets developers take components as-is, or they can extend a component's capabilities via inheritance. Developers create a new component by inheriting features and functions from existing components on the VCL, and change the components to meet the exact needs of their applications.

Creating components with Delphi is simply a matter of inheriting from an existing component type, defining new fields, properties, and methods, and registering the new component with Delphi. If you want to use third-party components (such as OCXs or VBXs) you need to register them with Delphi, so they will appear on the palette for use inside applications. The palette has a special panel for OCX controls, and Delphi automatically creates object wrappers for OCX controls so developers can use them as objects. This means that developers can create new hybrid objects from third-party components, and change object characteristics to meet application requirements. Notice how Delphi supports both Windows 3.1 and Windows 95 native control sets.

Like Delphi, Visual Basic lets developers assemble applications from a palette of components known as the toolbox. The toolbox is found on Visual Basic's Integrated Development Environment (IDE), where developers drag and drop components or controls onto waiting forms. Visual Basic uses several independent applications running concurrently to make up the development environment.

Also like Delphi, Visual Basic's toolbox is extendible through the integration of third-party OCXs. However, most developers will be happy with the components that come with Visual Basic. Once a control exists on a Visual Basic form, you can alter the properties of that control using the Properties window or Properties pages. You also have the option to set the properties of many controls at the same time using a grouping mechanism of Visual Basic.

Power Objects uses components similar to Visual Basic. Developers can load third-party OCXs into Power Objects for inclusion on the tool palette. Keep in mind that the lack of support for VBXs may mean that you will have a tough time finding third-party components for Power Objects until more VBX vendors release OCX versions of their products.

Inside the Power Objects development environment, libraries appear in the tool as book icons, and organize the globally accessed components. To place components on forms, you select a control on a palette and drop it onto the form. As with Visual Basic, you can set properties using a properties dialog.

In fact, if you're good at Visual Basic, you'll have no problem moving around Power Objects.

There is one area in which Power Objects differs greatly from Visual Basic: its support of the object-oriented development model. Power Objects bases its object on forms rather than on controls. Developers can make any form into a class and drop these classes onto other forms, thus inheriting the original form's characteristics. This is handy for applications that contain similar forms. When a developer makes changes to a form, the changes automatically propagate down to forms that inherit from that form. This is standard object-oriented functionality found in most client/server development tools.

All of these tools provide some sort of reporting capabilities. Visual Basic bundles Crystal Reports, and can add it into the tools and application as an OCX. Delphi bundles Borland's ReportSmith. Both of these commercially available report writers offer most of the reporting and business graphing capabilities you need for building complex applications. Power Objects, however, uses its own band-style report writer. This poses few problems with banded reports, but is somewhat awkward when creating complex reports. If you go with Power Objects, you should think about using another report writer (such as Crystal Reports, which integrates into Power Objects as an OCX).

Regarding the programming languages, you'll find that Visual Basic and Power Objects use an almost identical version of Basic. (Power Objects' version is known as Oracle Basic.) Both Basic languages are easy to use and understand. Delphi, unlike other tools on the market, uses Object Pascal. If you've taken Pascal in college, you'll find Object Pascal not much different. However, if you are learning from scratch, using a 3GL is not always desirable.

**Database Support**

All three tools can access a number of database engines, both local and remote. However, each approaches database access in its own unique way.

Delphi's Data Access pages provide Delphi objects for database access. For instance, you'll find the database objects that you need to connect to a database engine, access the tables, and query the database on behalf of the application.

The Borland Database Engine (BDE) is at the core of Delphi's database architecture. Delphi uses the BDE, including native database drivers, to access information on Oracle, Sybase, Informix, Interbase, DB2, and Microsoft SQL Server databases. You can also get to database files such as dBASE and Paradox through the ODBC standard interface. Delphi 2.0 has encapsulated much of BDE into high-level objects that you can edit visually in the application design environment. Of the three products, Delphi gives developers the most flexibility for database access. New with Delphi 2.0 is a database design facility that lets developers create and modify databases without having to use a third-party tool. If you need to browse and edit domains, triggers, and stored procedures, Delphi provides SQL Explorer. However, it is available only in the Client/Server Edition.

Visual Basic database access is built on top of the Data Access Object (DAO) and the JET (Joint Engine Technology) database engine. In addition, Visual Basic provides other database features, such as enhanced data control, new bound controls, and a database forms generator. Visual Basic's Database Manager offers complete control over any database engine that Visual Basic can connect to through 32-bit ODBC.

Through Visual Basic, JET supports the creation, modification, and deletion of tables, indexes, and database queries. Visual Basic and JET coordinate database access for Visual Basic applications using DAOs. DAOs provide developers with the ability to manipulate information in the database using the properties mechanisms of Visual Basic. Although data access from Visual Basic always seemed to be an afterthought, with version 4.0 it's clear that Visual Basic is finally serious about client/server development.

In addition to typical database operations, Visual Basic and JET provide special features such as data replication. This enables the developer to make entire copies of the database, and adjust for data and schema differences on the fly. Security is a feature as well, and JET allows the entire database to be password protected.

Power Objects is clearly the product of a database server company: ODBC is nowhere to be found. Instead, Power Objects uses native database drivers for Oracle, Sybase, and Microsoft SQL Server. Therefore, if your database is not among the privileged few, you're out of luck. However, Power Objects insists on complete control of the database engine, something that ODBC has difficulty with. Fortunately, ODBC support will be a part of the next release of Power Objects.

Power Objects pushes all database requests through a session object. The session object creates the connection to the database and visually places all accessible tables, views, and indexes on the workspace. Different icons represent different types of database resources. This visual representation of database resources is easy to use, more so than the other two.

After creating your database links, you can drag tables and views onto forms. This creates a bound record source between session objects and Power Objects containers. Most of the Power Objects controls can bind to the database, including list boxes, pop-up lists, and radio buttons.

Other database features of Power Objects include the ability to toggle locks from optimistic to pessimistic, and developers can control buffers for better integrity control. Also, Power Objects, like Visual Basic, provides the ability to move data and schemas from one database to another. This is Power Objects' "upsizing" strategy.

All three products offer local database engines, so developers don't have to hook up their applications to full-blown database servers to get started. Delphi provides a Windows version of Interbase, Power Objects provides Oracle's Blaze database, and Visual Basic provides local file access using ODBC. However, Blaze can't handle compound indexes.

**OLE Automation Fever**

A common theme among these three tools is support for OLE automation, both as clients and servers. Simply put, this is the ability of one application to provide services to another application through a common interface. For instance, using OLE automation, Visual Basic can use Microsoft Access as a report server, or you can send the results of a database query from Delphi to Microsoft Excel.

Visual Basic is one of the few tools that can extend the capabilities of OLE automation across a network. Using remote OLE automation, Visual Basic developers create a multi-tiered application with portions of an application running on remote servers as OLE automation servers. This enables developers to "partition" an application on as many application servers as required to meet the scaling requirements of the application. Although remote OLE automation provides an advanced application distribution mechanism for Visual Basic, it's difficult to set up. In reality, remote OLE automation is a precursor to Microsoft's upcoming Network OLE. If you want to use OLE automation for partitioning purposes, you may want to wait.

You will find support for OLE automation in Delphi 2.0. Delphi uses a new type called "variant" to provide transparent integration to OLE automation servers and controllers (clients). Using Delphi, you create OLE automation servers by defining an object that inherits from a type known as "TAutoObject." Next, you define some code with a new automated reserved word to expose properties and methods to other OLE-enabled applications. The identifier lets you expose properties, parameters, and function results using various types. At this point, you can create both in-process or out-of-process incarnations of an OLE automation server. If you find all of this confusing, you'll be happy to know that Delphi includes an expert that makes it easy to create OLE automation servers.

Visual Basic is the newest old hand at OLE automation. Creating an OLE automation server using Visual Basic is simply a matter of adding a new class module to the project for object definitions. Then you define the methods and properties for the server, set the "Class Public" property to True, and classify the Project StartMode as an OLE Server.

Unfortunately, Power Objects can act only as an OLE automation container. The ability to generate OLE automation servers is in the works. For now, this is the domain of Visual Basic and Delphi, or for those willing to traverse the obstacles of using a 3GL such as Visual C++.

## Performance

You can break down performance for these tools into development tool performance, runtime application performance, and database performance. All of the development tools perform well on my Windows 95-based Pentium 100 with 16MB of memory. However, on slower 486s, you will notice that Visual Basic begins to lag behind Power Objects and Delphi. The reality is that all that OCX stuff built into Visual Basic means it will get along better with faster PCs.

Windows 95 with Win32 support for Delphi and Visual Basic is essential for the developer's workstation. Although both of these tools can generate 16-bit (Win16) applications for Windows 3.1, these tools exploit the multi-tasking and multi-threading capabilities of the Windows 95 and Windows NT operating systems, and can generate applications that also exploit these capabilities. Power Objects and Power Objects-based applications, in contrast, are more at home on Windows 3.11, but run fine on Windows 95.

In terms of runtime performance, there is a clear winner. Delphi generates native .exes for application distribution, while Power Objects and Visual Basic are limited by p-code interpreters. This means that not only do Delphi applications perform better, they require fewer memory chips and processor cycles. Power Objects comes in a close second, with Visual Basic right on its heels.

All of these tools provide good database performance. However, Visual Basic's insistence on ODBC does mean that your database performance will lag behind Delphi and Power Objects, both of which use native database drivers. The inclusion of 32-bit ODBC almost makes up for the additional overhead of a least-common-denominator translation layer. Delphi seems to beat the others, although it was difficult for me to determine, because the test applications ran faster, and thus could draw the data to the screen quickly. The other tools could very well have the data back from the database engine, but application performance under the interpreters delays the time it takes to move the information to a screen or report.

**Picking Your Tools**

The right tool for you depends on your application requirements: platform considerations, ease of development, performance, and scalability. If you need Windows 95 or Windows NT 32-bit compatibility, your choices are limited to Visual Basic and Delphi. Both provide Win32 support, including support for multi-threading and multi-tasking. However, if you need to move your application between the Windows environment and the Macintosh, Power Objects is your only choice.

All of these tools enable developers to build applications quickly and with little effort. Visual Basic just barely surpasses Delphi and Power Objects with its quick ability to get applications up and running. Delphi's development environment is powerful, but it is a tad more complex to deal with as a novice. Power Objects is also a competent development environment, but it has a few kinks, such as the lack of a static menu, toolbar, and status line objects, as well as a limited number of OCXs. Both Visual Basic and Delphi are rich in components, and they have lucrative third-party native component markets that enable developers to create applications through integration, not perspiration.

If performance is your most important application requirement, then your choice is Delphi, hands down. The application deployment mechanism is the best in the business, and it lets you place true native executables on desktops rather than relying on p-code and interpreters. Borland seems to be keeping this trick to itself. The only way to match the performance of a Delphi application is to use a native 3GL and a compiler.

If you want scalability, then Visual Basic and Delphi are the contenders. Visual Basic's remote OLE automation servers are the first to offer easy application partitioning (multi-tier development) using OLE. However, you should know that this is not the Network OLE that Microsoft will eventually deliver. You should also know that this is bleeding-edge stuff, and it's going to take some time and trouble to build and set up remote OLE automation servers. Delphi provides you with the ability to create OLE automation servers with a migration path to the Microsoft OLE infrastructure when it finally hits the streets.

The good news is that all three of these competitors plan major upgrades in the near future, and you can count on great things from each. The bad news is that you'll have to sift through the new capabilities to find what your application really needs.

David S. Linthicum is a technical manager with EDS in Falls Church, Virginia. He's the author of several books on software development, a speaker, and an associate professor of computer science at a local college. You can email David at 70742.3165@compuserve.com, or visit his home page on the World Wide Web at http://ourworld.compuserve.com:80/home-pages/D_Linthicum/.

Reprinted with permission from DBMS Magazine, April, 1996, Vol. 9, No. 4, Copyright (c) 1996 Miller Freeman, Inc., ALL RIGHTS RESERVED.

**Summary**

Programming against the Win32 API is an exhaustive subject. There are many fine books on the subject. One of my favorites is Win32 System Services by Marshall Brain, published by Prentice Hall. This chapter served to provide you with an overview of the services available to programmers building robust and powerful Windows Telephony applications. Without all these services, WOSA libraries like MAPI and DAO, many TAPI applications would be featureless. In other words TAPI is not a self-contained environment, it is just one of the many books in the library of Windows functionality. The next two chapters are devoted entirely to TAPI functionality and programming.

# Chapter 4

# Windows Telephony and TAPI

This Chapter Covers

- The Evolution of Windows
- What Does TAPI Add to Windows?
- TAPI 101: An Introduction to the Telephony API
- Client/Server Telephony and TAPI 2.0

You have probably read a lot of definitions and a number of articles describing Microsoft's Telephony API. Somewhere in the Win32 API documentation you will read something similar to the following: The Microsoft Windows Telephony application programming interface (TAPI) provides services that enable a software developer to add telephone communications to applications developed for the Windows 16-bit and 32-bit operating systems

TAPI is a child (one of many) of the marriage between Intel and Microsoft. It is not really clear who really chipped the first sparks that led to this fiery API. Back in 1992 and 1993 I corresponded with several people at Intel, and at that stage it seemed that a lot of the initial ground work was being done by Intel. They were at the time working on a number of modems and fax cards which were to be TAPI compliant, but no serious TAPI-compliant line of hardware emerged for a while, not until sometime after the API was out of BETA and released in its first version.

## The Evolution of Windows

TAPI is one the many components of the Windows operating system that stands testimony to the marvelous evolutionary process taking place in the world of personal computing. When the PC was first introduced it was hard to use and scary for many. Then in the late 1980s Windows arrived. At first it was considered the computer industry's bad Joke. I remember the hated Unrecoverable Application Errors (UAEs) in the early versions, like

Windows 286; but I realized in these early years that we were standing on the cusp of a revolution in information technology; and besides, who could afford a $10,000 Apple system in 1989/90.

At first many feared that the Windows graphical user interface would be a drag and flunky. It *was* for a number of years because the pace of hardware innovation was slow and the prices were still very high. But as things changed and Window 3.00 arrived we began to use the computer as a personal productivity tool. We got stuck into spreadsheets, word processors, and database applications. Them came the desktop publishing systems (which wiped out the jobs of thousands of typesetters and paste-up people). Finally with the advent of Windows for Workgroups and the likes of the Lantastic product from Artisoft Corp, many small companies could afford a small peer to peer network. We had entered the age of workgroup computing.

LANs helped bring on the great email era. At that time the computer telephony industry was lighting its fires in the DOS, OS/2 and UNIX world. Windows in the meantime was fast become a pervasive operating system, rapidly taking over every desktop. IBM and Apple took a nap and woke up to find their market share abducted. Darwin called such an evolutionary process selective adaptation.

It comes thus as no surprise that the next logical step taken in the evolutionary process of the Windows operating systems is the combining of mankind's most useful tool, the telephone, with the computer. This maturing of Windows has led to the creation of the standard telephony interface for Windows, the Telephony Application Programming Interface (TAPI).

## What Does TAPI Add to Windows?

The following is an extract from one of the original TAPI documents released by Intel and Microsoft:

"Visual Interface to Telephone Features. Users can easily access telephone features through the Windows graphical user interface. Telephone providers have been adding features to telephones, including call forwarding, call parking, speed dialing, and conference calling. Unfortunately, most people typically use only one or two of the most basic features simply because they can't remember the more sophisticated features - let alone how to use them. Through telephony software applications, users will be able to access even the most complex features with ease. For example, setting up a conference call may be as easy as "clicking" the names of the conferees.

Personal Communication Management. With a graphical user interface, people can have their PCs handle incoming telephone calls, automatically controlling which calls reach them. For example, people will be able to ensure that they receive important calls by requesting that certain calls be forwarded automatically to locations at which they expect to be working.

Telephone Network Access. Connecting existing applications directly to the telephone network will greatly expand the usefulness of these applications and increase users' productivity. For example, a personal information manager can be used not only to look up a telephone number, but also to actually place a call to that number.

Additionally, several interesting new applications will be enabled by combining computers and telephones:

Integrated Messaging. With direct connection to the telephone, integrated messaging becomes a reality. Today, when people arrive at the office, they must check three separate "in-boxes": voice mail, electronic mail, and the fax machine. With the desktop computer connected to the phone network, users can combine these three forms of messaging into a single in-box. They can review and manage all of their messages—electronic mail, fax, and voice mail—-from a single place. In addition, voice-mail messages can be accessed randomly, which is far more efficient than the serial access provided on most telephone-based voice-mail systems.

Integrated Meetings. One of the most attractive capabilities of the computer is that it can store, communicate and present information that spans the entire spectrum of media—text, data, graphics, voice, and video—in any combination. By itself, the telephone can communicate and present voice information only. By combining the functions of the computer and the telephone, people in geographically separated locations can participate in interactive meetings and share visual as well as audio information. That means they can hold meetings over the telephone network that are nearly as rich in information content as in-person meetings."

Of course this did not all happen overnight. It's been a painful process. In particular the TAPI technology has been strapped by the confines of 16-bit programming. With the facilities of a full-blown 32-bit operating system, many routines were extremely difficult to implement. The operating systems were also far too fragile for the core of mission critical applications. Much of what we discussed in the previous chapter did not exist a year or two ago.

TAPI functions allow your applications to control telephony functions, services and routines. In other words your applications can call basic telephony functions, such as going off hook, to establish or answer calls and abandon or terminate calls. You can also access the standard switching functions of PBX systems and central exchange systems of the telephone company. Such as call holding, parking, transferring and conferencing.

There is also a service provider interface to the telephony API as illustrated in Figure 4-1. This allows service providers, the manufacturers of the telephone equipment, to extend services that are specific to their products. A good example is the Caller ID service that is finding its way onto many components. If a service provider's earlier versions of the equipment or component do not support Caller ID, TAPI will inform the application (in the form of an error message) as it attempts to hook into such functions.

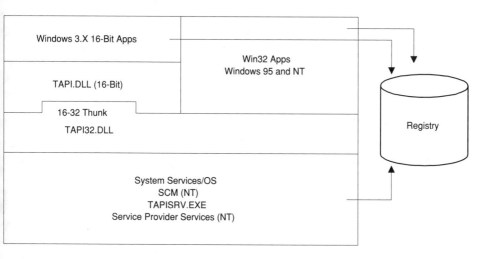

Figure 4-1: TAPI Architecture

The Windows Telephony API operates independently of the underlying telephone network and equipment, as described in Chapter 1. It essentially isolates you, the programmer, from the network, providing instead an abstraction above the network, a higher level access. This serves to hide the underlying complexity, which makes developing telephony applications a lot simpler. In many cases, however, you will still need to access manufacturer specific APIs and "talk" or "drill" down to specific equipment directly. TAPI does not hide everything, nor can it, nor should it.

The API also does not concern itself with how you connect to or install the telephony equipment. As illustrated in accompanying illustrations in Figure 4-2. You can a) connect directly to the PSTN with the use of a line interface card; or b) the PC can connect to the network via a modem or telephone; or c) it can be logically connected to the network via the interface provided by a telephony server.

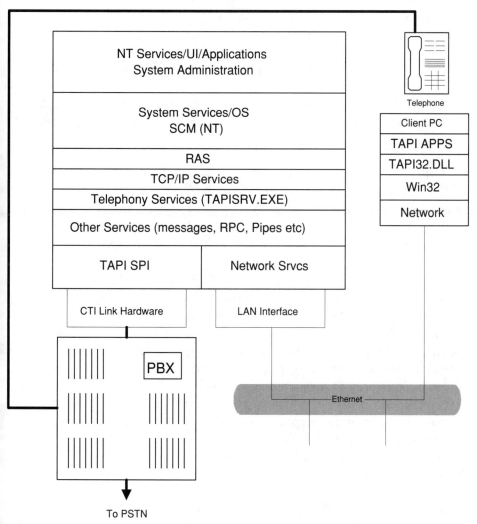

**Figure 4-2:** TAPI does not concern itself with how PCs and telephones connect to physical telephony devices

The Telephony API is part of the Win16 and Win32 SDKs (TAPI Versions 1.X through 2.0 across both the 16-bit and 32-bit API sets) is comprised of several core components:

1. The TAPI.DLL. Everything is centered around the TAPI.DLL, the libarary which provides the interface to the TAPI functions; thus facilitating a "dialog" between applications and service providers. Applications "talk" to the TAPI.DLL they never access the service provider directly. Under Windows 95 TAPI calls were translated from 32-bit code to 16-bit code via the services of the TAPI32.DLL.

2. The TAPI.LIB, provides entry points to the functions of TAPI.DLL.

3. TAPIEXE.EXE. This is an application that is automatically loaded by TAPI.DLL. It provides an application context for handling the telephony events.

4. TAPI.H. This is the header file for the API functions, which must be liked into the Windows telephony applications.

5. Also included is a number-translation module called TAPIADDR.DLL, which translates addresses from canonical to dialable format.

A new component has been added to TAPI, which plays an important role in the 32-bit TAPI 2.0 environment. This is the application service executable called TAPISRV.EXE. As of this writing it has been implemented in Windows NT 4.0, and should be available for Windows 95 by the end of 1996. According to the documentation "It essentially provides an execution context for the TAPI dynamic link library to carry out duties such as allocating memory or accessing data structures that have been swapped out to disk. This internal, system-only application runs in memory when running telephony applications, but does not create a window or appear in the task list (in other words it is not a user accessible item). The TAPI dynamic link library automatically loads and unloads the application as needed."

As an API for defining first-party call control, TAPI has solicited a great deal of support for applications that automate the dialing process, especially from smaller businesses. Microsoft was content, it seems, to leave TAPI on Windows clients, satisfying its commitment to provide an API for first-party call control. But according to Microsoft, it "discovered" that Windows NT provided a suitable platform for telephony, computer telephone integration and third-party call control on the server, a la TSAPI style. Hence, TAPI Version 2.0 for Windows NT (and future versions of Windows 95) was born. TAPI 2.0 is discussed later on in this chapter.

# TAPI 101: An Introduction to the Telephony API

When the Telephony Applications Programming Interface (TAPI) was first released to software developers in 1993 it classified two levels of service programmers could write for: Assisted Telephony and Telephony

**Assisted Telephony**

This first simple level provides telephony-enabled applications (such as contact managers and simple dialers) with a simple route to basic telephony functions that facilitate the making of phone calls. These tasks are so straightforward and basic that they do not require advanced knowledge of telephony. You simply call the functions to access the telephone network and establish a call. If that is all you need then you need to look no further than this part to enable your software. Anything more complex requires you to access the full API. The following code examples illustrate the extent to which you need to go to code dialer support into your application.

**Telephony**

The advanced second level provides access to three classes of telephony: Basic Telephony, Supplementary Telephony, and Extended Telephony. Let's begin our discussion with basic telephony.

Basic Telephony was the first and lowest level of telephony service that was initially defined in the API. It provided what the "Wintel" lads called a "guaranteed" set of functions that correspond to POTS (plain old telephone service) the simplest telephone service. In other words, the API on its release day only gave you direct (first-party) control of a telephone device (such as a modem in a PC) to make and receive calls. As a result many felt that Version 1.x of the API was not only line- and phone-centric, but confined to simple client applications.

However, the line- and phone-centric functions in question offered by the API were not only the most basic of telephony functions, they were, in my opinion also the most essential functions for the overwhelming majority of telephony transactions. And I believe they still are for the large majority of businesses in the world. Remember what I said about telephony and CTI back in the first two chapters. Many millions of businesses in the world are saddled with equipment that cannot be integrated with their computers systems or put under computer control without in-band signaling and communication. That means the only apps that work with these devices are the ones that can directly control hardware (phone devices) that "talk" to the switch or PBX over the plain old analog telephone line. Of course, it's not ideal to have a computer environment only partly enabled by TAPI, but at least its something (and all things taken into consideration, such as cost of components, adapters, and so on).

At the start of the any POT communication initiation is the logical line device. The logical line device is nothing more than the telephone line. (The TAPI terminology can be a little confusing at times because so many entities, and or attributes of the telephony electronics and equipment, are referred to as devices. It's hard to imagine a line as being a device. Most refer to the line device as a "twisted pair.") But before we get any further into understanding what the logical line device in the API means for our telephony application, let's try and understand basic telephony, ala TAPI, a little more. If you skipped the first two chapters you may wish to go over them now because it will help you make sense of certain programming solutions provided by the API.

The following functions are included in Basic Services as listed in the API documentation:

### Fundamental API interactions

- Initialization and shutdown of the telephony API
- Negotiation of API version
- Filtering of status messages to be received
- Opening and closing line devices

### Capabilities and Status

- Querying line device and address capabilities, associated media stream devices, icons, etc.
- Querying line device, address, and call status
- Querying and setting media device configurations

### Operations

- Translating addresses and setting associated parameters
- Dialing and answering calls
- Dropping calls
- Controlling call privileges and handing off call control

## Supplementary Services

The Supplementary Telephony service provides (in-band) access to switch features, such as hold, transfer, conference, and so on. It essentially includes all the services in the API that allow you to access the features found on all modern PBXs.

These supplementary features are considered optional, which means that you cannot assume that all service provider equipment will support any or all of a particular feature or function. For example, Park, Orbit, or Camp-on-hold may all seem to mean the same thing, but depending on whose equipment you are integrating with very different things take place inside the switch.

To determine the capabilities you have to code the application to query the service provider equipment (using functions such as lineGetDevCaps or lineGetAddressCaps, see below).

The following is a summary of the functions included in Supplementary Services, as exhibited in the TAPI documentation:

- Holding and unholding calls
- Transferring calls, both blind and with consultation (see Chapter 1 and 2)
- Conferencing
- Forwarding calls
- Parking and unparking calls
- Picking up calls that are ringing elsewhere
- Camp-on and other automatic call completion
- Accepting, rejecting, and redirecting incoming calls
- Securing calls from interruptions
- Generating in-band dial digits and tones (see Chapter 2)
- Monitoring media mode, received DTMF digits, and tones
- Controlling the routing of media stream information
- Sending user-user information (ISDN)
- Changing call parameters on-the-fly
- Controlling the physical phone terminal (speakers, microphones, ringers, display, lamps, buttons, etc.)

## Extended Services

The Extended Telephony level provides numerous and well-defined API extension mechanisms that enable application developers to access service-provider-specific functions not directly defined by TAPI. (TAPI defines the extension mechanism only, and thus the definition of the Extended-Service behavior must be completely specified by the service provider.)

The extended services provide for some exciting possibilities, however I am yet to see many service providers taking full advantage of TAPI's extension mechanism. The service-providers can define new values for some of the enumeration types and bit flags, and to add fields to most of the data structures. Special functions and messages such as **lineDevSpecific** and **phoneDevSpecific** are provided in the API to allow an application to directly communicate with a service provider. The parameters for each function are also defined by the service provider.

## Telephony Devices

### A Line Device

To best understand the idea of a line device you should refer back to Chapter 1 and understand the working of the local loop, loop current, and the public switched telephone network or PSTN. From the TAPI programming viewpoint a line device needs to be qualified in terms of its capabilities. For example a line device could be assumed to be a plain old analog service or it can be an ISDN line that has numerous digital features. At the physical layer (down to the plain old copper wires) the lines are no different. What determines the difference is whether the line is transmitting signals (as audio waves) or digital information (collections of 1s and 0s).

### A Phone Device

When is a phone a phone device. Seems redundant, does it not. A phone device is any device that provides an implementation for the phone-prefixed functions in the Telephony API. As with the Line device class, the phone device is an abstraction of a physical phone. In that the API treats line and phone devices as devices that are independent of each other, your applications can use a phone without having to use an associated line. You can also access a line without using a phone. This means that it is possible to hook into the phone device and extrapolate its usage into other services, such as using the telephone as an audio input device/interface to the computer.

### Address Formats and Address Translation

The typical method of dialing people or fax machines or Internet hubs is by selecting the number from an address book. Address book management is a nightmarish situation. How, for example, do applications know to append or pre-pend digits, depending on the circumstances surround the apparent call. And how does the dialer know when a number is for a fax machine, or a telephone or both. Both geographic and location information needs to be supplied in some situations. TAPI has an address translation scheme that lets the user inform the computer of the geographic location and the desired line device needed for the call. TAPI then automatically handles the dialing differences, requiring no changes to the user's address book. You need to call the lineTranslateAddress function to convert an address from the canonical address format to the dialable address format.

What is the so-called canonical address. For starters it includes the seemingly out-of-place "+" character, then the country code, city or area code surrounded by parentheses (2711), and the local number consisting of the exchange digits (555) and the extension hanging off the exchange (1212). The canonical address is the essential string needed to inform the central office and the PSTN of the number the application is trying to reach. You can then append other digits, such as voice mail boxes and passwords, to the end of the canonical address. The TAPI canonical address format is similar to the international format for printed telephone numbers defined by the (CCITT). (See Appendix A). So it looks like the following:

+ (1212) 555 1212,,,345

The string to be dialed consists of digits and characters that need to be translated in the appropriate in-band tones recognized by the interfacing telephony system. These include characters for switch or PSTN services such as the hookflash; switching between pulse and tone dialing; pausing; waiting for a second dial tone (such as the dial-tone provided by the PSTN, see chapters 1 and 2) and so on. The calling card "bong" needs to also be taken into consideration. Of course, although you can use the "dialable" addresses returned by **lineTranslateAddress**, your application it is not limited to such a string. You can offer up any string and fire a dialing sequence.

## Call Handles and Call Privileges

The TAPI documentation describes a call as a representation of a connection between two (or more) addresses: "The originating address (the caller) is the address from which the call originates, and the destination address (the called) identifies the remote end point, station or remote phone device with which the originator wishes to communicate."

TAPI identifies a specific call by means of the call's handle. The so-called handle is not something specific to TAPI. The concept of establishing handles to "objects" is universal throughout the Win32 API. You typically have to establish a single call handle for every call apparent to the calling (as in function call) application. Calling TAPI call functions results in the creation of new call objects. These objects return the new call's handle to the application. Often call handles will materialize outside the bounds of the code you use to establish calls. These are typically the handles that TAPI creates as a result of inbound call detection data being passed through the interface layers to the TAPI.DLL.

In TAPI philosophy calls can "accessed" by one or more applications which establish an "ownership" privilege over the call. These applications are allowed to manipulate and influence the call state via a handle to the call. The application's privilege on the call is classed as being either an owner or a monitor. An application that does not have ownership privilege over a call (but has a handle to it) is a monitor of the call and is prevented from manipulating it. As a monitor of the call the application is allowed to query the calls state and extrude information about the call as necessary to achieve certain results. More than one application can have call monitoring privilege over a TAPI call.

How does your application gain ownership of a call. First, if it creates the call; that is, establishing a call for outbound dialing it by default gains exclusive ownership of the call. Ownership of the inbound calls is assigned by TAPI according to application specific needs and configuration.

TAPI permits applications that have a "monitoring handle" to switch to an "ownership handle." This nifty feature is known as "hand off." Essentially it means that the original owner of the call can hand off ownership of the call to another application. This handing off of a call is a feature allowed between TAPI applications. Both service providers and the underlying equipment and physical environment remain oblivious to this juggling of handles by the application or applications.

The primary reason for the various levels of direct control services mentioned above is that TAPI Version 1.X primarily concerned itself with first-party call control (see the Lexicon). In other words, it only provided for direct control of a telephone device (and thus the call) if that device was installed on the data bus of the computer running a Windows client (3.1 and 95) or via a connector (such as a dialer connection to the parallel port). This method is the converse of the logical connection, the keystone of Novell and AT&T's Netware Telephony Services (See Computer Telephony Strategies, Jeffrey R. Shapiro, IDG Books Worldwide Inc., 1996).

## Client/Server Telephony and TAPI 2.0

TAPI 2.0 provides third-party call control and client/server computer telephony for both Windows 95 and Windows NT. This service will serve the objective of providing a CTI connection between the telephony enabled computer and the switch. However, TAPI is no longer just an external API or SDK. It is now an integral part of the Win32 API; which lies at the very core of Windows NT and Windows 95. (It is also implemented within the object oriented OLE/Common Object Model (COM) vision, context and philosophy of Microsoft, DEC and others, allowing users to collect families of telephony components that interoperate with each other on a computer or across a network . . . . but that's a subject for the next edition of this book.)

With TAPI supported on both the client and the server sides, Windows can provide a rich telephony-services environment under the 32-bit operating systems. TAPI 2.0 essentially defines services that enable Windows NT and Windows 95 to function as both a telephony clients and telephony servers. For example, you will be able to run voice-processing cards, fax modems, and switch cards in the server, and, on the same machine, and maintain a CTI link to a switch. This CTI link will provide the logical connection between devices on users' desktops (digital or analog telephones connected only to the switch).

Under Win32 system services, applications are exposed to the power of 32-bit OS wonders such as symmetrical multiprocessing (SMP—in Chapter 3), preemptive multitasking and multithreading. With the release of TAPI 2.0 it is now be possible to provide third party call control power and functionality typical of formal and informal call center applications (any busy office for that matter), as separate processes of the Win32 system architecture.

Understand that call center and third party call control applications are not new to the Windows operating systems. Many applications have been "running" on Windows for some time now, but only accessing standard OS services. However, with switch, telephony and voice processing companies providing (service provider) support, in the form of native VxD drivers and TAPI service provider software, call control functionality becomes native to system services (a facility of the OS, rather than stuff built into application internals that are too critical to be considered "safe" so high above the OS code base).

This means that the switch, a PBX, ACD or some queuing device, needs never concern itself directly with the application running on the client Windows computer; it only needs to "talk" to the TAPI service on the server.

As long as the client or service requester application (such as a Windows NT voice mail server) is TAPI 2.0 compliant it will be able to request services from the switch, such as checking to see if a station (someone's telephone is off-hook). Computer telephony applications thus need only make requests to the Win32 API for service, which assumes the job of receiving and servicing the requests from application software and the telephony devices at the equipment service provider end.

There is support for station status, ACD call queuing; routing, transferring and conferencing; outbound call processing and predictive dialing; agent monitoring and control (call data feeds); call state and event timers and control (see below). TAPI 2.0 also provides support for what Microsoft calls "quality of service parameters," which refers to catering to the advent of new carrier technologies such as Asynchronous Transfer Mode (ATM).

TAPI was first implemented as a standard 16-bit dynamic link library or DLL on Windows 3.X. It did not ship as part of the operating system; and vendors needed to specifically ship the DLL to end users with their application. Then as Windows 95 became a reality the API was simply migrated to the new 32-bit operating system with no "native" 32-bit capability. It was simply distributed along with the operating system for the benefit of Windows 95 communications utilities.

Under Windows 95, 32-bit applications simply made their requests to the 16-bit API through the process known as "thunking" which is "simply" the translation of 32-bit code into 16-bit code. Microsoft's road to a full 32-bit API was never really clear. Nor was the journey and the estimated time of arrival something the independent software developers could bank on, nor would they have wanted to.

In order to develop a robust, full featured and advanced computer telephony system—preferably as a family of business components that individual companies *can* best leverage and manipulate to suite their specific business purpose and for their reasons only—you will require a lot more than TAPI. I have made this point earlier, in other books and in several articles. Many other Win32 system services and Win 32 APIs are needed to program against for features such as remote procedure calls (RPC), access to message service providers, database management and more.

By fully integrating telephony services into the Win32 API, Microsoft has made Windows a true telephony and computer telephony operating system. In the remainder of this chapter I have tried to provide a reference that does a better job of just repeating or re-publishing the TAPI documentation that ships with the Win32 API and the TAPI 2.0 SDK. I had wanted to fully understand the new TAPI 2.0 and all its functions for this book, but I received the pre-release API too late, even though it was not yet formerly shipping to any developers nor was it available in the SDK or the Microsoft Developer Network, to commit to a comprehensive full study. And at this writing few developers are shipping systems that are fully TAPI 2.0 enabled. Again, we will leave this to the next edition and I will try my best to provide you with a general overview of the new 32-bit architecture. Where I deemed it best to leave TAPI 2.0 documentation in its original text I have enclosed the passage in quotation marks.

## TAPI 2.0 Features:

**Native 32-bit Support**

All the core TAPI components are now fully implemented as 32-bit components. This means that the developers can develop to the TAPI with the full offering of the Win32 API, and seize the ability to code against Microsoft's most powerful components as part of their computer telephony and Windows telephony objectives. Thus you can develop Windows telephony applications that take full advantage of the multiple platforms (Windows NT), SMP, preemptive multitasking, and thread technology (see Chapter 3), and, of course, OLE technology and all it offers and promises.

**16-bit/32-bit application portability**

If you have existing 16-bit TAPI applications which currently run on Windows 95 and Windows 3.1 (using the TAPI 1.3 API) they will be able to run on Windows NT without modification or recompilation. If you have applications that have been implemented in Win32 architectures, such as applications running on Windows 95 and developed against the TAPI 1.4 API, they will be able to run as is on Windows NT Intel machines without the necessity to recompile the executable.

**The New 32-bit Service Provider Interface**

Under TAPI 16-bit Windows 3.X applications called functionality in the 16-

bit TAPI.DLL. The service provider or application provider usually shipped the DLL and installed it into the Windows System directory (or folder in Windows 95). Before the advent of TAPI 2.0 existing 32-bit applications operating on Windows 95 and Windows NT merely linked to the thunk layers to enable them to perform telephony function on the Win32 operating systems.

In TAPI 2.0 the core or central component of TAPI is an executable service called TAPISRV.EXE. You can see it loaded and started in Windows NT via the Server Manager application or via the Diagnostics application. TAPISRV executes as a "service process" and replaces the TAPIEXE.EXE hidden process and the TAPI.DLL.

Telephony Service Providers will create services that will execute in TAPISRV.EXE's process. This new architecture will eliminate the difficulties that materialized in previous versions of Windows Telephony which were caused by service providers inadvertently executing in the TAPI.EXE context. This led to serious problems in that separate processes could destroy the telephony resources or objects owned by processing applications.

Underneath the service provider DLL (TSP), the service provider can use any system functions or other components necessary, including **CreateFile** and **DeviceIOControl** to work with IHV-designed kernel mode components and services as well as standard devices such as serial and parallel ports to control external locally-attached devices. They can also access network services (such as RPC, WinSockets, and Named Pipes) for client-server telephony, as outlined in Chapter 3.

In addition to the new "installed" service you also have two new telephony objects that can exist in a Win32 TAPI system. The first of these objects is called the Telephony Service Provider User Interface DLL. This DLL is loaded by TAPI into the process of an application that invokes any of the service provider functions that can display a dialog (for example, **lineConfigDialog**). The service provider can also cause its associated UI DLL to be loaded and executed in the process of an application if the service provider needs to display dialogs at "unexpected" times (for example, the "Talk/Hang-up" dialog displayed by the Universal Modem Driver when a data modem is used to dial an Interactive Voice call using **lineMakeCall**. The **lineMakeCall** function is not normally considered to be a UI-generating function).

The second new component type is known generically as a Proxy Request

Handler. This is a Full Telephony application that normally executes on a telephony server (the same server on which the telephony service provider is executing for the associated line device(s)). This architecture, rather than the WOSA "service provider" architecture, is used when a particular service is more appropriately implemented in an application than a driver on the server. In this version of TAPI, the ACD Agent management functions would be implemented in a proxy request handler rather than in a service provider.

The Universal Modem Driver (UNIMODEM) service provider is also available on Windows NT for modem control. Windows Telephony for Windows NT also includes a generic kernel mode Telephony Service Provider Interface (TSPI) mapper, known as KMDDSP, that allows service providers to be implemented as kernel mode device drivers.

Telephony Service Providers on Windows NT must be Win32; 16 bit service providers will not run on Windows NT under TAPI 2.0. A 16-bit compatibility layer will be provided on the next major release of the Windows 95 platform (I won't hazard a guess, but I believe this is scheduled for the end of 1996.) Existing 16-bit Windows 3.1 and Windows 95 service providers can continue to operate on upgraded Windows 95 systems.

The service provider consists of a minimum of two components. The main service provider DLL (designated in the Figure 4-1 as "MAIN.TSP") executes in the context of the TAPI service, and performs all of the tasks of the service provider that are not related to user interface components associated with a particular application's use of the device (most likely, in conjunction with lower-level components not shown in the diagram). But unlike previous versions of TAPI in which the user interface code was integrated into the main service provider (and executed, because of the previous architecture, within the context of the application), the service provider must now include a separate component that implements the user interface elements.

## TAPI 2.0 Specific Components

### Unicode

Win32 applications can choose to call the existing ANSI TAPI functions or to call Unicode versions of functions that pass or return strings (functions with a "W" suffix). A TAPI 2.0 service provider must be fully Unicode; all strings passed as parameters to TSPI functions or returned by service providers in structures or through pointers must be in Unicode.

## Service processes

Inherent in TAPI 2.0 are mechanisms that allow you to create TAPI enabled services processes in the service facility of Window NT. The service facility is one of NT's most powerful features. The service concept on Windows NT is similar to the "daemon" concept in UNIX, only a lot more powerful. For example, an SMTP service running in Windows can pump email out to the Internet at rates that leave UNIX daemons choking. These services that can interface with applications just like any service process. This means you can fire off, respond to, or trigger telephony events, or notify applications of telephony events, without a Window message queue. RPC and TCP services can be TAPI enabled. The telephony service processes can operate in the background and comfortably execute TAPI services, such as keeping track of station status information (see the section on Services in the previous chapter).

## NDISTAPI compatibility

The NDISTAPI defines how WAN miniport NIC drivers implement telephonic services. (The Win32 SDK includes an overview of how a TAPI-capable WAN miniport initializes, brings up lines, sets up calls and closes calls using TAPI and WAN OIDs (object identifiers.) What this means is that the existing support in Windows NT 3.5 for ISDN WAN miniports under Remote Access Service will be preserved. NDIS WAN miniport drivers will be supported under a kernel mode service provider without modification.

## Registry support

The registry replaces the old Windows initialization file architecture, which was insecure, especially for telephony functions. Now all telephony parameters are written to and read from the registry, which typically is only accessed by the applications and service processes. The registry concept is implemented in both Windows 95 and Windows NT. Under Windows NT registry objects can be tightly secured, making it difficult to tamper with telephony setting and parameters. Telephony service providers and all stored parameters can be updated across the LAN.

## Call Center Support

The TAPI 2.0 documentation talks about the concept of "modeling a call center." This means that enhancements have been made to TAPI and the TSPI to support the functionality required in a call center environment, formal and informal. TAPI 2.0 thus enables you to create predictive dialing applications and applications that control and manage call queues, ACD systems, station status management and station synchronization.

The following text segment was extracted from the TAPI 2.0 document which illustrates the new call center capability: "Service providers can expose each resource on the PBX as a line device and possibly an associated phone device. Terminals which support multiple call appearances would do so through multiple addresses, just as in first party call control. In fact, the third party view of a device is identical to the first party view; applications on the server can see and control all of the "first party" devices, whereas an individual client PC connected to the server would only be able to see those devices which are made visible to it though access controls administered by TAPISRV.EXE on the server (the presumption is that the granularity of security for devices exposed by the server would be lines and phones, rather than addresses or calls). Resources other than terminals can also be modeled as line devices. For example, an ACD queue or route point would be modeled as a line device that could have many active calls; an IVR server, voicemail server, or set of predictive dialing ports could also be modeled as a line device that supports multiple calls.

Within this model, the status of the addressed device and calls associated with it can be monitored though existing TAPI messages such as **LINE_LINEDEVSTATE, LINE_ADDRESSSTATE, LINE_CALLSTATE, and LINE_CALLINFO,** and details obtained through functions such as **lineGetLineDevStatus, lineGetAddressStatus, lineGetCallStatus,** and **lineGetCallInfo.** Whenever a TAPI object is operated upon through a "third party" application running on the server, the result is identical to what would have occurred if the same object had been similarly operated on by a "first party" application running on a client PC associated with that device. Status indications sent by the server service provider controlling the switching fabric (or switch) are delivered both to applications running on the server and to those running on associated, authorized clients."

This support is probably one of the most exciting in terms of the development of extensive and extensible computer telephony server systems, such as PBX systems, ACD and, especially, voice and fax processing systems. It means that powerful server products can be created to provide systems that can deliver computer telephony services and applications to all clients without the client's need to host expensive equipment locally. Modems, fax cards, and especially voice processing cards, are expensive items. Even the cheapest Rhetorex card is still around $230.00. With the support mentioned above it is possible to create a fully TAPI enabled voice processing system-cum-PBX without any hardware needed in client computers. Again if you read Chapter 2 you will note the significance of the ability to monitor device and call states via the existing TAPI messages such as **LINE_LINEDEVSTATE, LINE_ADDRESSSTATE, LINE_CALLSTATE, and LINE_CALLINFO.**

"Predictive dialing" as mentioned earlier is an application that typically runs on the telephony server. It access a a list of phone numbers from a database and attempts to make outbound calls. At a certain predetermined point in the completion of the call (possibly when a human says "hello") the call is automatically assigned to a call center position for processing and handling. The application may make use of a "predictive dialing port" on the telephony server, often a voice processing card, which can make outbound calls and has the necessary technology (such as DSP) to detect call progress tones and other call state information.

The following text from the TAPI 2.0 documentation illustrates the possibilities:

Predictive dialing capability is established "by a new **LINEADDRCAPFLAGS_PREDICTIVEDIALER** bit in **LINEADDRESSCAPS.dwAddr CapFlags.** A new field in **LINEADDRESSCAPS, dwPredictiveAuto TransferStates,** indicates the states upon which the predictive dialing port can be commanded to automatically transfer a call; if this field is zero, it indicates that automatic transfer is not available, and that it is the responsibility of the application to transfer calls explicitly upon detecting the appropriate call state (or media mode or other criteria). Preferably, switch manufacturers will make available both automatic and "manual" transfer, and allow applications to select the preferred mechanism, but service providers would have to model the behavior of legacy equipment. A single predictive dialing port (line device/address) may support making several outbound calls simultaneously, as indicated by **LINEADDRESSCAPS.dwMaxNumActiveCalls.**

"Predictive dialing capability can also be made available on any device, using a shared pool of predictive dialing signal processors, which are bridged onto the line being dialed upon request. When the **lineMakeCall** function is used on a line capable of predictive dialing (a port with the **LINEADDRCAPFLAGS_PREDICTIVEDIALER** set) and predictive dialing is requested using **LINECALLPARAMFLAGS_PREDICTIVEDIAL**, then the call is made in a predictive fashion with enhanced audible call progress detection. Additional fields and constants are defined in the **LINECALLPARAMS** structure passed in to **lineMakeCall** to control the behavior of the predictive dialing port. The new field **dwPredictiveAutoTransferStates** indicates the **LINECALLSTATES** which, upon entry of the call into any of them, the predictive dialing port should automatically transfer the call to the designated target (the bits must be a proper subset of the supported auto-transfer states indicated in **LINEADDRESSCAPS**); the application can leave the field set to 0 if it desires to monitor call states itself and use **lineBlindTransfer** to transfer the call when it reaches the desired condition. The application must specify the desired address to which the call should be automatically transferred in the variable field defined by the new **LINECALLPARAMSdwTargetAddressSize** and **dwTargetAddressOffset.**

Applications can also set a time-out for outbound calls so that the service provider will automatically transition them to a disconnected state if they are not answered. This is controlled using the **dwNoAnswerTimeout** field in **LINECALLPARAMS.**"

Call Queues and Route Points

The concept of a call queue and a route point is also new to TAPI, and presents the opportunity to create exciting applications. The call queue or route point is a special address within the switch where calls are temporarily held pending an action or event. "This characteristic is represented by new bits in **LINEADDRESSCAPS.dwAddrCapFlags, LINEADDRCAPFLAGS_QUEUE** and **LINEADDRCAPFLAGS_ROUTEPOINT.** All calls appearing on such an address are awaiting action by the application, and there may be default actions that take place (such as, transfer to an agent or trunk) if the application takes no action within a defined period of time. The application must be configured by the system administrator so that it knows what actions it should take regarding calls appearing on each queue or route point address, and the amount of time available to decide on the action to take.

"Applications can determine the number of calls pending in a queue or route point using **lineGetAddressStatus. lineGetCallInfo** can be used to obtain information such as calling ID, called ID, inbound or outbound origin, etc., and used by the application to make decisions on call handling; calls can be redirected, blind-transferred, dropped, etc., or just allowed to automatically pass out of the queue to a destination. A call goes to **LINECALLSTATE_DISCONNECTED** if it is abandoned. Calls go **IDLE** when they leave the queue; **lineGetCallInfo** can be used to read the redirection ID to determine where they were transferred.

"Some switches allow calls in a queue or on hold to receive particular treatment such as silence, ringback, busy signal, music, or listening to a recorded announcement. A new TAPI function **lineSetCallTreatment** allows the application to control the treatment. The structure delimited by **LINEADDRESSCAPS.dwCallTreatmentListSize** and **dwCallTreatmentListOffset** allows applications to determine the supported treatments. **LINECALLINFO.dwCallTreatment** indicates the current treatment, and a **LINE_CALLINFO** message with **LINECALLINFOSTATE_TREATMENT** indicates when this changes. The new **LINECALLFEATURE_SETTREATMENT** bit in **LINECALLSTATUS.dwCallFeatures** indicates when changing the treatment by the application is permitted. The **LINECALLTREATMENT_** set of constants defines a limited set of predefined call treatments; service providers may define many more."

**ACD Agent Monitoring and Control**

The monitoring and control of ACD agent station status is supported through the addition of seven new TAPI functions:

lineGetAgentCaps
lineGetAgentStatus
lineGetAgentGroupList
lineGetAgentActivityList
lineSetAgentGroup
lineSetAgentState
lineSetAgentActivity

And a new **LINE_AGENTSTATUS** message is used to indicate when agent information has changed.

"Note that these controls are associated with an ADDRESS instead of a LINE because many ACD systems are implemented with different ACD queues associated with buttons on the phone terminal (and separate call appearances). Also, ACD agent phones may often have separate call appearances for personal calls. Architecturally, we prefer ACD functionality to be implemented in a server-based application. The client functions mentioned above, rather than mapping to the telephony service provider, are conveyed to a server application which has registered (using an option of lineOpen) as a handler for such functions. The new **LINE_PROXYREQUEST** message is used to signal to the handler application when a request has been made; it calls the **lineProxyResponse** function to return results and data. Handler apps can also call **lineProxyMessage** to generate **LINE_AGENTSTATUS** messages when required. In the case of a legacy PBX or standalone ACD which implements ACD functionality itself, the telephony service provider for the switch must include a proxy server application that accepts the requests and routes them (possibly using **lineDevSpecific** functions or a private interface) to the service provider, which routes them to the switch.

## Call Data

In a call center environment, applications may need to accumulate data about a call (such as IVR input of account numbers) that is desirable to have available to all agents and applications that handle the call. TAPI previously did not have a method for a telephony application to provide to the service provider data that could be passed along when a call was transferred and made visible to other applications that are monitoring the call (either on the same PC, or, through the server, on other PCs). Win32 TAPI has a new variable-sized field in the **LINECALLINFO** instructure, bounded by **dwCallDataSize** and **dwCallDataOffset,** to contain such call data. A new **LINECALLINFOSTATE_CALLDATA** message indicates whenever this field changes. A new function, **lineSetCallData**, allows an application that owns the call to set this data; **LINECALLFEATURE_SETCALLDATA** indicates when changing the data is permitted. **LINEADDRESSCAPS. dwMaxCallData** indicates the maximum number of bytes permitted in this field. Initial call data to be attached to a call can be passed to the service provider in **LINECALLPARAMS.**

## Station Status Control

There are three major station status functions that need control: message waiting lights, forwarding, and do not disturb. Forwarding and Do Not Disturb are controllable through the existing **lineForward** function (which is address-specific), and queried using **lineGetAddressStatus**. The **LINEDEVSTATUSFLAGS_MSGWAIT** bit in **LINEDEVSTATUS. dwDevStatusFlags** indicates the status of the message waiting light on the device, and a **LINEDEVSTATE_MSGWAITON** or **LINEDEVSTATE_ MSGWAITOFF** message is sent to indicate when the state changes. All that is needed is a function to allow the message waiting light to be controlled without having to implement a TAPI Phone device just for that purpose. The new function lineSetLineDevStatus is defined for this purpose; the **LINEFEATURE_SETDEVSTATUS** bit (in **LINEDEVCAPS.dwLine Features** and **LINEDEVSTATUS.dwLineFeatures**) indicates when it can be called, and **LINEDEVCAPS.dwSettableDevStatus** allows the application to detect which of the device status settings can be controlled from the application. In addition to allowing the message waiting feature to be controlled, it also allows the device's Connected, Inservice, and Locked status to be set, to the extent that these are supported by the switch or other hardware. Calls to this function result in appropriate **LINEDEVSTATE** messages being sent to reflect the new status.

**Call State Timer**

Currently, all timing of calls is left up to applications. This can be quite burdensome if the application is monitoring a large number of calls, and if multiple applications were present, possibly on multiple servers, it would be necessary for them to all maintain timers on the same calls. It therefore makes more sense for call state timing to be handled by the server. A new field is added to **LINECALLSTATUS** to allow timing of calls in states to be reported. **LINECALLSTATUS.tStateEntryTime** (of type **SYSTIME**) indicates the time at which the current state was entered.

**Quality of Service Support**

As Asynchronous Transfer Mode (ATM) networking emerges into the mainstream of computing, and support for ATM is added to other parts of Windows NT, TAPI must also be extended to support key attributes of establishing calls on ATM facilities. The most important of these from an application perspective is the ability to request, negotiate, renegotiate, and receive indications of Quality of Service parameters on inbound and outbound calls. By the way QOS is not restricted to an ATM transport provider; any service provider can implement QOS features.

TAPI now supports the ability to monitor the quality of service available to applications accessing telephone and data networks servicing the enterprise. "Applications will be able to request, negotiate, and renegotiate quality of service (performance) parameters with the network, and receive indication of QOS on inbound calls and when QOS is changed by the network. The Quality of Service structures are binary-compatible with those used in the Windows Sockets 2.0 specification." Quality of service information is accessed from the underlying digital transport mechanism that can notify applications when the quality of service is better or worse on additional channels. It is worthwhile to read the Windows Sockets 2.0 specification. Quality of service specifics play and important role in the nature and use of sockets. For example the quality of service requirements for a video conference will be a lot stricter than for a simple voice conversation.

According to the new documentation Quality of Service information in TAPI is exchanged between applications and service providers in **FLOWSPEC** data structures which are defined in Windows Sockets 2.0. "Applications request QOS on outbound calls by setting values in the **FLOWSPEC** fields in the **LINECALLPARAMS** structure. The service provider will endeavor to provide the specified QOS, and fail the call if it cannot; the application can then adjust its parameters and try the call again. Once a call is established, an application can use the lineSetQualityOfService function to request a change in the QOS; a new bit, **LINECALLFEATURE_SETQOS**, lets applications determine when this function may be called. The QOS applicable to inbound or active calls can be obtained by using lineGetCallInfo and examining the new **FLOWSPEC** fields. A new bit in the **LINE_CALLINFO** message, **LINECALLINFOSTATE_QOS**, lets applications know when QOS information for a call has been updated."

### Enhanced Device Sharing

TAPI 2.0 enables applications to "restrict their handling of inbound calls on a device to a single address, to support features such as distinctive ringing when used to indicate the expected media mode of inbound calls." This means it is possible to programmatically notify applications (and humans) if an inbound call is meant for a fax machine, a modem or a telephone set. Applications making outbound calls can also set the device configuration when dialing out.

### User and Kernel Mode Components

Under Windows NT TAPI services and processes, including top-level service provider DLLs, can run in both user and kernel mode. This means that if your application requires access to certain kernel mode components (which may be the case in comprehensive voice servers system architecture) the TAPI support can be implemented in Kernel mode. TAPI code that runs in the kernel mode is given controlled access (by NT Executive) to system hardware and information. This ability to process in the either kernel mode or user mode is essential architecture for establishing mission critical service architectures. (See Chapter 3 for the appropriate discussion of Windows NT architecture.)

### Media Event Timing

Media event time is also new to TAPI. It is possible to write computer telephony code that acts on a timing relationship between media events, such as the interval between DTMF digits received. A good example of this is in the automated attendant feature of a system where a long pause between digits routes a caller to a menu option, such as "press 1 for" and where digits pressed in quick succession are translated into extensions requested.

With media event timing it is now possible to timestamp a media event to determine the relative timing between events. The timestamps use the millisecond-resolution as "time since Windows started." The time is returned to the application by the **GetTickCount** function.

The TAPI messages that can be timestamped are **LINE_GATHERDIGITS, LINE_GENERATE, LINE_MONITORDIGITS, LINE_MONITORMEDIA,** and **LINE_MONITORTONE**. The tick count gets inserted into the dwParam3 of these messages. According to the documentation "if timestamping is not supported by the service provider (which is indicated by the service provider setting dwParam3 in these messages to 0), then TAPI itself will insert the tick count into dwParam3 of all of these messages (it may be skewed somewhat, but less than if the application did the same after the messages had traversed an interprocess communication scheme)."

**Summary:**

This chapter provided an overview of the TAPI architecture and an overview of both basic and extended telephony services. The following chapter provides and in-depth discussion of the most important functions in the TAPI library and an overview of the TAPI programming model.

## Chapter 5
# Key TAPI Functionality

This chapter provides a basic understanding of the TAPI programming model and serves as an introduction to the key TAPI functions. It will provide you with a little more (I hope) of an idea of what it takes to create an application to set up the basic outbound telephony services . . . initialize TAPI and create outbound calls.

It is important to understand that TAPI has succeeded in abstracting the complexities of telephony and the underlying telephony network and devices from the programmer. While this does not mean that you should ignore the workings of the telephone system (which will not be useful), the philosophy espoused by the TAPI creators is such that they have created the tools for you to cater to these complexities by calling TAPI to handle the underlying complexity in your stead. Without TAPI you would have to obtain the C interface library to every device your program will encounter. . . what a drag. With TAPI as the universal telephony translator your applications simply call on the services of TAPI to perform telephonic miracles in software. This approach allows you to concentrate more on services and features than on technology.

Rather than provide you with 250 pages of application source code to hack, I deemed it better to provide information that will assist you to determine if the native third generation languages, C/C++ coding or Pascal/Object Pascal, calls against the TAPI is the best route for your purposes. You may for example be better off working at higher levels above the API by using TAPI components, such as the toolkits offered by Stylus Innovation in their Visual Voice product line. I have annotated some Delphi source code in Appendix B to provide an introduction to this as well.

Your application, however, may require you to access TAPI directly. And an extensive computer telephony application may need the services of both native API access and higher level components. Meanwhile, the TAPI 32 documentation in the TAPI SDK and several references, such as the

Microsoft Developer Network (MDN) include all the functions for TAPI applications, and you should refer to these works for an extensive study or investigation into the entire library of functions.

Following is a list of the application interface API function calls and constants supported by TAPI. As you will notice many are suited to client/server telephony software, such as an ACD system. Items referenced with (2.0) indicate the function is new and applicable to TAPI version 2.0. and, is thus, specific to client/server telephony. Note too that this list is not exhaustive and does not contain information indicating whether the functions complete asynchronously or synchronously. (For a complete list of API messages, data structures and constants refer to the Win32 API documentation.)

## API Functions

| Function | Purpose |
| --- | --- |
| lineAddProvider | Used to add a service provider |
| lineAgentSpecific | new |
| lineAnswer | Used to answers an inbound call |
| lineClose | Closes a specified opened line device |
| lineCompleteTransfer | Used to complete transfer on colsultation |
| lineConfigDialog | Displays a configuration dialog box to enable a user to configure or setup a device |
| lineConfigDialogEdit | Same functionality as above. Implemented in Version 1.4 |
| lineConfigProvider | modified |
| lineDeallocateCall | Deallocates the specified call handle |
| lineDial | Dials a specified or converted string |
| lineDrop | Abandons a call |
| lineGetAddressCaps | Returns the telephony capabilities of an address |
| lineGetAddressStatus | Returns current status of a specified address |
| lineGetAddressID | Retrieves the address ID of an address specified using an alternate format |
| lineGetAgentActivityList | Returns a list of the activities performed by an agent (2.0) |
| lineGetAgentCaps | Returns the capabilities of an agent, such as ACD group participation (2.0) |
| lineGetAgentGroupList | New, identifies agent groups |
| lineGetAgentStatus | New, this function identifies the current status of an agent at a particular address |
| lineGetAppPriority | modified |
| lineGetCallInfo | Returns constant information about a call |
| lineGetCallStatus | Returns complete call status information for the specified call |

| Function | Purpose |
|---|---|
| lineGetConfRelatedCalls | Returns a list of call handles that are part of the same conference call as the call specified as a parameter |
| lineGetCountry | Retrieves dialing rules and other information about a given country. Added in version 1.4 |
| lineGetDevCaps | Returns the capabilities of a given line device |
| lineGetDevConfig | Returns configuration of a media stream device that can be used by the application |
| lineGetIcon | Used by the application to display an icon to the user |
| lineGetID | Retrieves a device ID associated with the specified open line, address, or call. Modified for version 2.0 |
| lineGetLineDevStatus | Returns current status of the specified open line device |
| lineGetMessage | Returns the next TAPI message that is queued for delivery to an application that is using the Event Handle notification mechanism(2.0) |
| lineGetNewCalls | Returns call handles to calls on a specified line or address for which the application does not yet have handles |
| lineGetNumRings rings set | This function returns the minimum number of see also lineSetNumRing below |
| lineGetStatusMessages | Returns the application's current line and address status message settings |
| lineGetTranslateCaps | Returns address translation capabilities |
| lineHandoff | Hands off call ownership or changes an application's privileges to a call, or both |
| lineInitialize | Obsolete in 2.0, see LineInitializeEx |
| lineInitializeEx | Initializes the application's use of TAPI and returns the number of line devices available to the application requesting service (2.0) |
| lineMakeCall | Establishes an outbound call and returns a call handle for it |
| lineNegotiateAPIVersion | Used by an application to negotiate the TAPI version |
| lineOpen | Open a specified line device for providing subsequent monitoring and/or control of the line |
| lineProxyMessage | Used by a registered proxy request handler to generate TAPI messages related to its activities. This function allows the owner of a handle to the line to generate proxy messages that can be received by all applications that have the line open (2.0) |

| Function | Purpose |
|---|---|
| lineProxyResponse | Indicates completion of a proxy request by a registered proxy handler such as an ACD agent handler on a server (2.0) |
| lineRemoveProvider | Modified in Version 2.0. See LineAddProvider |
| lineSetAgentActivity | Sets the agent activity parameters (2.0) |
| lineSetAgentGroup | Sets the agent groups into which the agent is logged into on a particular address (2.0) |
| lineSetAgentState | Sets the state of a particular agent. For example a logged out agent will not be passed new calls (2.0) |
| lineSetAppPriority | Modified in version 2.0 to permit long file names |
| lineSetCallData | Facilitates the ability to access data concerning a call. Applications that have the line open have access to the call data. |
| lineSetCallPrivilege | Sets the application's privilege over the call |
| lineSetCallQualityOfService | Allows the owner application to request quality of service of the service provider, or to change quality of service as needed (2.0) |
| lineSetCallTreatment | Allows the owner application to control (by requesting the service provider) what a caller hear while staying on-hold (2.0) |
| lineSetCurrentLocation | Sets the location used as the context for address translation. Used for area code ratification |
| lineSetDevConfig | Sets the configuration of the specified media stream device |
| lineSetLineDevStatus | Allows applications with handles to the line to get line status information from the service provider |
| lineSetNumRings | Sets the number of rings before inbound calls are to be answered |
| lineSetStatusMessages | Specifies the status changes desired by the application |
| lineSetTollList | Used to manage toll lists |
| lineTranslateAddress | Converts an address stored in canonical format to an address that can be dialed |
| lineTranslateDialog | Displays dialog box allowing to user to change location and calling card information. Added in Version 1.4 and modified in Version 2.0 |
| Phone functions | |
| phoneClose | Closes a phone device |
| phoneConfigDialog | modified |
| phoneGetIcon | modified |

| Function | Purpose |
|---|---|
| phoneGetMessage | Returns the next TAPI message that is queued for delivery to an application that is using the Event Handle notification mechanism (2.0) |
| phoneInitialize | Obsolete |
| phoneInitializeEx | Initializes the application's use of TAPI and returns the number of phone devices available to the application requesting service (2.0). (This function replaces phoneInitialize) |
| phoneOpen | Opens a phone device |
| phoneShutdown | Terminates the applications use of the phone device |

## Steps to Success

A TAPI application can be divided into three distinct components (I call them phases): Start-up, Management, and Tear-Down (clean-up). Like a three-legged stool the application will fall over if any one or more of the parts is missing or is unstable. Here we highlight the key start-up phase functions.

## Start-up Phase

There are three distinct actions that an application takes in the start-up phase. They are TAPI initialization, API version negotiation, and line opening (for dialing or inbound call handling). In the startup phase your program initializes TAPI and establishes the API version on the hosting computer.

Start-up phase also enables the API function callbacks and establishes information about the lines (line devices) connected to the computer. In many respects its a way of saying to the application before we take any actions let's see what telephony resources we have available. You can't begin communication without first determining whether the computer is connected to POTS, ISDN, and so on. Once this has been accomplished your application can begin communicating with TAPI to receive messages.

# Initializing TAPI: The LineInitializeEx Function

Function: LineInitializeEx

```
LONG lineInitializeEx
    (LPHLINEAPP lphLineApp,
     HINSTANCE hInstance,
     LINECALLBACK lpfnCallback,
     LPCSTR lpszFriendlyAppName,
     LPDWORD lpdwNumDevs,
     LPDWORD lpdwAPIVersion,
     LPLINEINITIALIZEEXPARAMS lpLineInitializeExParams);
```

Return Values: The function returns zero if the call succeeds.
It returns a negative error number if an error has occurred.
Return values are:
LINEERR_INVALAPPNAME
LINEERR_OPERATIONFAILED
LINEERR_INIFILECORRUPT
LINEERR_INVALPOINTER
LINEERR_REINIT
LINEERR_NOMEM
LINEERR_INVALPARAM

You have two functions with which to accomplish this task (neither acts on any hardware device, or initializes any physical component: **LineInitializeEx** and **phoneInitializeEx**. TAPI is not normally memory resident unless a specific service (typically on NT) initializes a telephone service at boot time and maintains its running state at all times. The first application to initialize TAPI invokes a telephony environment on the host machine. This environment consists of the loaded TAPI.DLL, TAPISRV.EXE and any supporting collateral device drivers and service provider components. The information is usually stored in the system registry.

API Notes: The functions **lineInitializeEx** and **phoneInitializeEx** are the only two callbacks defined in TAPI. This means that they are the only two functions that can pass the pointer of a callback function to TAPI. All subsequent TAPI functions, when completing asynchronously, result in calls to this callback function. (However, in Assisted Telephony, the functions **tapiRequestMediaCall** and **tapiRequestDrop** use window messages to return call states to the application that is placing the call (such as MS Word or the dialer services in Excel menus or ACT). All other functions that terminate asynchronously (see Lexicon) return calls to these functions. Assisted telephony apps, return information to the calling applications via message dialogs.

**Explanation of the Parameters and Usage**

The **lphLineApp** parameter specifies a pointer to a location that returns a structure containing the application's usage handle for TAPI. The application captures the handle and releases it when done with TAPI services. This parameter allows Windows to keep tabs on the applications using the telephony devices.

The **hInstance** parameter specifies the instance handle of the client application or DLL. The application or DLL may pass NULL for this parameter and not provide any instance information. TAPI will thus use the module handle of the root executable of the process which will be either an executable file or a DLL. (TAPI requires this information for purposes of identifying call handoff targets and media mode priorities). You would also typically use this parameter to keep track on the several instances of the process, which would be useful in event logging procedures.

The **lpfnCallback** parameter specifies the address of a callback function that is invoked to determine status and events on the line device, addresses, or calls, when the application is using the "hidden window" method of event notification (for more information see **lineCallbackFunc**). This parameter is ignored and should be set to NULL when the application chooses to use the "event handle" or "completion port" event notification mechanisms.

The **lpszFriendlyAppName** parameter specifies a pointer to a NULL-terminated ASCII string that contains only displayable ASCII characters. The string is used, if this parameter is not NULL, to return the name of the calling application. If it is NULL the application's file or module name will be used instead (such as telefone.exe or telefone.dll). This name is provided in the **LINECALLINFO** structure to indicate, in a user-friendly way, which application originated, or originally accepted or answered the call. This information can be useful for call logging purposes and debugging. You would typically call the Windows API function GetModuleFileName if the **lpszFriendlyAppName** structure is NULL; and the application's module file name would be used instead.

The **lpdwNumDevs** parameter specifies a pointer to a DWORD-sized location. Upon successful completion of this request this location is filled with the number of line devices available to the application.

The **lpdwAPIVersion** parameter specifies a pointer to a DWORD-sized location. The application must initialize this DWORD, before calling this function, to the highest API version it is designed to support (in other words, the same value it would pass into **dwAPIHighVersion** parameter of the **lineNegotiateAPIVersion** function). Artificially high values must not be used; the value must be accurately set (for this release, to 0x00020000 (TAPI 2.0)). TAPI will translate any newer messages or structures into values or formats supported by the application's version. Upon successful completion of this request, this structure is populated with the highest API version information supported by TAPI, thereby allowing the application to detect and adapt to having been installed on a system with an older version of TAPI.

The **lpLineInitializeExParams** parameter specifies a pointer to a structure of type **LINEINITIALIZEEXPARAMS** containing additional parameters used to establish the association between the application and TAPI.

Application design and logic will revolve around this parameter which governs how your application receives, interprets and acts on messages returned by TAPI. Service processes require very different message handling routines than GUI based applications. Applications that support the 32-bit architecture and thus TAPI version 2.0 have a choice of three message or event-notification schemes to deploy: Hidden windows, event handle, or completion port. The method you use is applied in the call to this function.

The hidden windows scheme calls on TAPI to create a sub-classed hidden window into which all messages due to your application are posted. Calls to the **GetMessage API** function streams the messages into a message queue which the application then services.

The second message processing option is the event handle. Here TAPI creates an application event object, which carries the message, and returns a handle to the object in the hEvent field in **LINEINITIALIZEEXPARAMS**.

You then call the **lineGetMessage** function to retrieve the message.

The third method is the completion port. Your application creates the completion port when initializing TAPI. All messages are then serviced at the completion port in a similar fashion to the hidden windows scheme.

The above function is the TAPI "ignition key" in a nutshell. Besides any other startup functions you wish to call, or sundry events that your application needs to trigger or perform in a start-up mode, this is about all it takes to place Windows into telephony mode.

## Negotiating the API Version

 **Function: LineNegotiateAPIVersion**

LONG lineNegotiateAPIVersion(
    HLINEAPP hLineApp,
    DWORD dwDeviceID,
    DWORD dwAPILowVersion,
    DWORD dwAPIHighVersion,
    LPDWORD lpdwAPIVersion,
    LPLINEEXTENSIONID lpExtensionID)

Return Values: The function returns zero if the call succeeds.
It returns a negative error number if an error has occurred.
Return values are:
LINEERR_BADDEVICEID
LINEERR_NODRIVER
LINEERR_INCOMPATIBLEAPIVERSION
LINEERR_OPERATIONFAILED
LINEERR_INVALAPPHANDLE
LINEERR_RESOURCEUNAVAIL
LINEERR_INVALPOINTER
LINEERR_UNINITIALIZED
LINEERR_NOMEM
LINEERR_OPERATIONUNAVAIL
LINEERR_NODEVICE

The idea behind this function (which is a concept that will soon be deployed across the entire Windows API) is to provide applications that have been written to older versions of TAPI with continuing support in the new library. In other words if a user installed a newer version of TAPI on his or her PC, the old application would continue to use the new TAPI version as if it was the older version. This is achieved by keeping the old or obsolete functions in place for so-called "backward compatibility." New functions in the new library thus become the preferred calls for new applications, but old applications still have access to obsolete functions. The **LineInitialize** function is an example; it has now been replaced with the **LineInitializeEx** example described above.

**Explanation of the Parameters and Usage**

The **hLineApp** parameter specifies the handle to the application's registration with TAPI.

The **dwDeviceID** parameter specifies the line device to be queried.

The **dwAPILowVersion** parameter specifies the least recent API version the installed application is compliant with. The high-order word is the major version number; the low-order word is the minor version number.

The **dwAPIHighVersion** parameter specifies the most recent API version the application is compliant with. The high-order word is the major version number; the low-order word is the minor version number.

The **lpdwAPIVersion** parameter specifies a pointer to a DWORD-sized location that contains the API version number the application negotiated. If the negotiation is successful, this number populating this structure will be in the range between **dwAPILowVersion** and **dwAPIHighVersion.**

The **lpExtensionID** parameter specifies a pointer to a structure of type **LINEEXTENSIONID**. If the service provider for the specified dwDeviceID supports provider-specific extensions, then, upon a successful negotiation, this structure is filled with the extension ID of these extensions. This structure contains all zeros if the line provides no extensions. An application can ignore the returned parameter if it does not use extensions, which would be the case in many simple dialer examples.

Use lineInitialize to determine the number of line devices present in the system. The device ID specified by **dwDeviceID** varies from zero to one less than the number of line devices present (for example 0 to 3 on a four line device).

The **lineNegotiateAPIVersion** function is used to negotiate the API version number to use. It also retrieves the extension ID supported by the line device, and it returns zeros if no extensions are supported. If the application wants to use the extensions defined by the returned extension ID, it must call **lineNegotiateExtVersion** to negotiate the extension version to use.

The API version number negotiated is that under which TAPI can operate. If the TAPI application is too old and the TAPI library too new and an error is returned indicating that the application and the API versions are incompatible.

## Opening a Line Device

This function opens the line device specified by the device ID. TAPI returns a handle to the line device that can be used for subsequent operations and control over the line device, such as setting call privileges.

## Explanation of the Parameters and Usage

> Return Values: The function returns zero if the call succeeds.
> It returns a negative error number if an error has occurred.
> Return values are:
> LINEERR_ALLOCATED; LINEERR_LINEMAPPERFAILED
> LINEERR_BADDEVICEID; LINEERR_NODRIVER
> LINEERR_INCOMPATIBLEAPIVERSION; LINEERR_NOMEM
> LINEERR_INCOMPATIBLEEXTVERSION
> LINEERR_OPERATIONFAILED
> LINEERR_INVALAPPHANDLE; LINEERR_RESOURCEUNAVAIL
> LINEERR_INVALMEDIAMODE; LINEERR_STRUCTURETOOSMALL
> LINEERR_INVALPOINTER; LINEERR_UNINITIALIZED
> LINEERR_INVALPRIVSELECT; LINEERR_REINIT
> LINEERR_NODEVICE   LINEERR_OPERATIONUNAVAIL

The **hLineApp** parameter specifies a handle to the application's registration with TAPI.

The **dwDeviceID** parameter identifies the line device to be opened. It can either be a valid device ID or the value: **LINEMAPPER**. This value is used to open a line device in the system that supports the properties specified in **lpCallParams**. The application can use **lineGetID** to determine the ID of the line device that was opened.

The parameter **lphLine** specifies a pointer to an **HLINE** handle, which is then loaded with the handle representing the opened line device. You then use this handle to identify the device when invoking other functions on the open line device.

The parameter **dwAPIVersion** specifies the API version number under which the application and Telephony API have agreed to operate. This number is obtained with **lineNegotiateAPIVersion** explained earlier.

The parameter **dwExtVersion** specifies the extension version number under which the application and the service provider agree to operate. This number is zero if the application does not use any extensions. This number is obtained with **lineNegotiateExtVersion.**

The parameter **dwCallbackInstance** specifies user-instance data passed back to the application's callback. This parameter is not interpreted by the Telephony API.

The parameter **dwPrivileges** specifies the privilege the application intends to claim over the calls it obtained notification for. This parameter can be a combination of the **LINECALLPRIVILEGE_** constants. For applications using API version 0x00020000 or greater, values for this parameter can also be combined with the **LINEOPENOPTION_ constants.**

**LINECALLPRIVILEGE_NONE.** The application needs to make only outbound calls.

**LINECALLPRIVILEGE_MONITOR.** The application needs to monitor inbound and outbound calls.

**LINECALLPRIVILEGE_OWNER.** The application needs to own inbound calls of the types specified in **dwMediaModes.**

**LINECALLPRIVILEGE_MONITOR** and **LINECALLPRIVILEGE_OWNER.** The application needs to own inbound calls of the types specified in dwMediaModes, but if it cannot be an owner of a call, it needs to be a call monitor.

**LINEOPENOPTION_SINGLEADDRESS.** The application is only concerned with new calls that appear on the address specified by the

**dwAddressID** field in the **LINECALLPARAMS** structure pointed to by the **lpCallParams** parameter (which must be specified). If **LINEOPENOPTION_SINGLEADDRESS** is specified but either **lpCallParams** is invalid or the included **dwAddressID** does not exist on the line, the open fails with **LINERR_INVALADDRESSID**. In addition to setting the **dwAddressID** member of the **LINECALLPARAMS** structure to the desired address, the application must also set **dwAddressMode** in **LINECALLPARAMS** to **LINEADDRESSMODE_ADDRESSID**.

The **LINEOPENOPTION_SINGLEADDRESS** option affects only TAPI's assignment of initial call ownership of calls created by the service provider using a **LINE_NEWCALL** message. An application that opens the line with **LINECALLPRIVILEGE_MONITOR** will continue to receive monitoring handles to all calls created on the line. Furthermore, the application is not restricted in any way from making calls or performing other operations that affect other addresses on the line opened.

The **LINEOPENOPTION_PROXY** concerns an application that is willing to handle certain types of requests from other applications that have the line open, as an adjunct to the service provider. Requests will be delivered to the application using **LINE_PROXYREQUEST** messages; the application responds to them by calling **lineProxyResponse**, and can also generate TAPI messages to other applications having the line open by calling **lineProxyMessage**.

When this option is specified, the application must also specify which specific proxy requests it is prepared to handle. It does so by passing, in the **lpCallParams** parameter, a pointer to a **LINECALLPARAMS** structure in which the **dwDevSpecificSize** and **dwDevSpecificOffset** members have been set to delimit an array of **DWORDs**. Each element of this array shall contain one of the **LINEPROXYREQUEST_** constants. For example, a proxy handler application that supports all five of the Agent-related functions would pass in an array of five **DWORDs** (dwDevSpecificSize would be 20 decimal) containing the five defined **LINEPROXYREQUEST_** values. The proxy request handler application can run on any machine that has authorization to control the line device. However, requests will always be routed through the server on which the service provider is executing that actually controls the line device. Thus, it is most efficient if the application handling proxy requests (such as ACD agent control) executes directly on the server along with the service provider. Subsequent attempts, by the same application or other applications, to open the line device and register to handle the same proxy requests as an application that is already registered fail with

**LINEERR_NOTREGISTERED.** To stop handling requests on the line, the application simply calls lineClose, which will handle the tear-down process.

Other flag combinations return the **LINEERR_INVALPRIVSELECT** error.

The **dwMediaModes** parameters specify the media mode(s) of concern to the application. The **dwMediaModes** parameter is used to register the application as a potential target for inbound call and call handoff for the specified media mode. This parameter is meaningful only if the bit **LINECALLPRIVILEGE_OWNER** in **dwPrivileges** is set (and ignored if it is not). This parameter uses the following **LINEMEDIAMODE_ constants:**

**LINEMEDIAMODE_UNKNOWN.** The application needs to handle calls of unknown media type (unclassified calls).

**LINEMEDIAMODE_INTERACTIVEVOICE.** The application needs to handle calls of the interactive voice media type. That is, it manages voice calls with the human user on this end of the call (such as voice messaging or audiotext).

**LINEMEDIAMODE_AUTOMATEDVOICE.** Voice has been detected on the call. The voice is locally handled by an automated application.

**LINEMEDIAMODE_DATAMODEM.** The application needs to handle calls with the data-modem media mode.

**LINEMEDIAMODE_G3FAX.** The application needs to handle group 3 fax calls.

**LINEMEDIAMODE_TDD.** The application needs to handle TDD (Telephony Devices for the Deaf) type calls.

**LINEMEDIAMODE_G4FAX.** The application needs to handle group 4 fax calls.

**LINEMEDIAMODE_DIGITALDATA.** The application needs to handle digital data calls.

**LINEMEDIAMODE_TELETEX.** The application needs to handle telextex calls.

**LINEMEDIAMODE_VIDEOTEX.** The application needs to handle video text calls.

**LINEMEDIAMODE_TELEX.** The application needs to handle telex calls.

**LINEMEDIAMODE_MIXED.** The application needs to handle ISDN mixed media mode calls.

**LINEMEDIAMODE_ADSI.** The application needs to handle calls with the ADSI (Analog Display Services Interface) media mode.

**LINEMEDIAMODE_VOICEVIEW.** The application needs to handle VoiceView calls.

The **lpCallParams** parameters specify pointers to a structure of type **LINECALLPARAMS.** This pointer is only used if **LINEMAPPER** is used . . . otherwise **lpCallParams** is ignored. It describes the call parameter that the line device should be able to provide.

Return Values: Returns zero if the request is successful or a negative error number if an error has occurred.

Possible return values are:
LINEERR_ALLOCATED
LINEERR_LINEMAPPERFAILED
LINEERR_BADDEVICEID
LINEERR_NODRIVER
LINEERR_INCOMPATIBLEAPIVERSION
LINEERR_NOMEM
LINEERR_INCOMPATIBLEEXTVERSION
LINEERR_OPERATIONFAILED
LINEERR_INVALAPPHANDLE
LINEERR_RESOURCEUNAVAIL
LINEERR_INVALMEDIAMODE
LINEERR_STRUCTURETOOSMALL
LINEERR_INVALPOINTER
LINEERR_UNINITIALIZED
LINEERR_INVALPRIVSELECT
LINEERR_REINIT
LINEERR_NODEVICE
LINEERR_OPERATIONUNAVAIL

If **LINEERR_ALLOCATED** is returned, the line cannot be opened due to a "persistent" condition, such as that of a serial port being exclusively opened by another process. If INEERR_RESOURCEUNAVAIL is returned, the line cannot be opened due to a dynamic resource over-commitment such as in DSP processor cycles or memory. This over-commitment may be transitory, caused by monitoring of media mode or tones, and changes in these activities by other applications may make it possible to reopen the line within a short time period.

If **LINEERR_REINIT** is returned and TAPI re-initialization has been requested, for example as a result of adding or removing a Telephony service provider, then lineOpen requests are rejected with this error until the last application shuts down its usage of the API (using **lineShutdown**), at which time the new configuration becomes effective and applications are once again permitted to call lineInitialize.

Opening a line always entitles the application to make calls on any address available on the line. The ability of the application to deal with inbound calls or to be the target of call handoffs on the line is determined by the **dwMediaModes** parameter. The lineOpen function registers the application as having an interest in monitoring calls or receiving ownership of calls that are of the specified media modes.

If the application just wants to monitor calls, then it can specify **LINECALLPRIVILEGE_MONITOR**. If the application just wants to make outbound calls, it can specify **LINECALLPRIVILEGE_NONE**. If the application is willing to control unclassified calls (calls of unknown media mode), it can specify **LINECALLPRIVILEGE_OWNER** and **LINMEDIAMMODE_UNKOWN**. Otherwise, the application should specify the media mode it is interested in handling.

The media modes specified with **lineOpen** add to the default value for the provider's media mode monitoring for initial inbound call type determination. The **lineMonitorMedia** function modifies the mask that controls **LINE_MONITORMEDIA** messages. If a line device is opened with owner privilege and an extension media mode is not registered, then the error **LINEERR_INVALMEDIAMODE** is returned.

An application that has successfully opened a line device can always initiate calls using **lineMakeCall, lineUnpark, linePickup, lineSetupConference** (with a **NULL hCall**), as well as use **lineForward** (assuming that doing so is allowed by the device capabilities, line state, and so on).

Note that a single application may specify multiple flags simultaneously to handle multiple media modes. Conflicts may arise if multiple applications open the same line device for the same media mode. These conflicts are resolved by a priority scheme in which the user assigns relative priorities to the applications. Only the highest priority application for a given media mode will ever receive ownership (unsolicited) of a call of that media mode. Ownership can be received when an inbound call first arrives or when a call is handed off.

Note that any application (including any lower priority application) can always acquire ownership with **lineGetNewCalls** or **lineGetConfRelatedCalls**. If an application opens a line for monitoring at a time that calls exist on the line, **LINE_CALLSTATE** messages for those existing calls are not automatically generated to the new monitoring application. The application can query the number of current calls on the line to determine how many calls exist, and, if it wants, it can call **lineGetNewCalls** to obtain handles to these calls.

An application that handles automated voice should also select the interactive voice open mode and be assigned the lowest priority for interactive voice. The reason for this is that service providers will report all voice media modes as interactive voice. If media mode determination is not performed by the application for the **UNKNOWN** media type, and no interactive voice application has opened the line device, voice calls would be unable to reach the automated voice application, but be dropped instead.

The same application, or different instantiations of the same application, may open the same line multiple times with the same or different parameters. When an application opens a line device it must specify the negotiated API version and, if it wants to use the line's extensions, it should specify the line's device-specific extension version. These version numbers should have been obtained with **lineNegotiateAPIVersion** and **lineNegotiateExtVersion**.

Version numbering allows the mixing and matching of different application versions with different API versions and service provider versions.

The **LINEMAPPER** allows an application to select a line indirectly-by means of the services it wants from it. When opening a line device using **LINEMAPPER**, the following becomes applicable: All fields from beginning of the **LINECALLPARAMS** data structure through **dwAddressMode** are relevant. If **dwAddressMode** is LINEADDRESSMODE_ADDRESSID it means that any address on the line is acceptable, otherwise if dwAddressMode is LINEADDRESSMODE_DIALABLEADDR, indicating that a specific originating address (phone number) is searched for, or if it is a provider-specific extension, then **dwOrigAddressSize/Offset** and the portion of the variable part they refer to are also relevant. If **dwAddressMode** is a provider-specific extension additional information may be contained in the **dwDeviceSpecificvariably** sized field.

## Translating an Address String to Dial

Function: LineTranslateAddress

```
LONG lineTranslateAddress(
    HLINEAPP hLineApp,
    DWORD dwDeviceID,
    DWORD dwAPIVersion,
    LPCSTR lpszAddressIn,
    DWORD dwCard,
    DWORD dwTranslateOptions,
    LPLINETRANSLATEOUTPUT lpTranslateOutput)
```

Return Values: The function returns zero if the call succeeds.
It returns a negative error number if an error has occurred.
Return values are:
LINEERR_BADDEVICEID; LINEERR_INVALPOINTER
LINEERR_INCOMPATIBLEAPIVERSION; LINEERR_NODRIVER
LINEERR_INIFILECORRUPT;     LINEERR_NOMEM
LINEERR_INVALADDRESS;       LINEERR_OPERATIONFAILED
LINEERR_INVALAPPHANDLE; LINEERR_RESOURCEUNAVAIL
LINEERR_INVALCARD; LINEERR_STRUCTURETOOSMALL
LINEERR_INVALPARAM

This function lets you convert canonical address information in a string of digits that a modem or telephony device can dial. It converts canonical information into dialable information. The function has grown from a few parameters you could pass with version 1 to calling card and phone card support. Most telephone numbers or numerical addresses are stored in forms that make it easier for humans to remember and recall from directory information. For example, it is far easier to read and dial 1 (800) 555-1212 than 18002451212. We automatically understand that we must dial a "1" digit, the 800 number area code, and then the number to dial to reach toll-free calling directory information (in the U.S.). But a dialer or software does not "see it that way." It dials only the digits in the form of a string usually supplied to an alpha-numeric field. The dialer cannot dial the parenthesis or any dashes, so it needs to convert the telephone number into a "dialable" address string. This is referred to as the translated result string.

## Explanation of the Parameters and Usage

The **hLineApp** parameter specifies the application handle returned by **lineInitializeEx**.

The **dwDeviceID** parameter specifies the device ID for the line device upon which the call is intended to be dialed, so that variations in dialing procedures on different lines can be applied to the translation process. In other words a device that needs to dial access digits may behave differently to a device directly connected to the exchange, possibly using a different transmission media (both such devices can thus exist in the same system).

The **dwAPIVersion** parameter returns the highest version of TAPI supported by your application (not necessarily the value negotiated by **lineNegotiateAPIVersion** on some particular line device).

The **lpszAddressIn** specifies a pointer to a NULL-terminated ASCII string containing the address from which the information is to be extracted for translation. This information must be in either the canonical address format, or an arbitrary string of dialable digits (non-canonical). This parameter must not be NULL. If the **AddressIn** contains a subaddress or name field, or additional addresses separated from the first address by ASCII CR and LF characters, only the first address is translated, and the remainder of the string is returned to the application without modification.

The **dwCard** specifies a credit card to be used for dialing. This field is only valid if the **ARDOVERRIDE** bit is set in **dwTranslateOptions**. This field specifies the permanent ID of a Card entry in the [Cards] section in the registry (as obtained from **lineTranslateCaps**) which should be used instead of the **PreferredCardID** specified in the definition of the **CurrentLocation**. It does not cause the **PreferredCardID** parameter of the **current Location** entry in the registry to be modified; the override applies only to the current translation operation. This field is ignored if the **CARDOVERRIDE** bit is not set in **dwTranslateOptions** parameter below.

The **dwTranslateOptions** specifies the associated operations to be performed prior to the translation of the address into a dialable string. This parameter uses the following **LINETRANSLATEOPTION_** constants: **LINETRANSLATEOPTION_CARDOVERRIDE**. If this bit is set, **dwCard** specifies the permanent ID of a Card entry in the [Cards] section in the registry (as obtained from lineTranslateCaps) which should be used instead of the **PreferredCardID** specified in the definition of the CurrentLocation. It does not cause the **PreferredCardID** parameter of the current Location entry in the registry to be modified; the override applies only to the current translation operation. The **dwCard** field is ignored if the **CARDOVERRIDE** bit is not set.

**LINETRANSLATEOPTION_CANCELCALLWAITING:** If a "Cancel Call Waiting" string is defined for the location, setting this bit will cause that string to be inserted at the beginning of the dialable string. This is commonly used by modem and fax applications to prevent interruption of calls by call waiting beeps (intrusion). If no Cancel Call Waiting string is defined for the location, this bit has no affect. Note that applications using this bit are advised to also set the **LINECALLPARAMFLAGS_SECURE** bit in the **dwCallParamFlags** field of the **LINECALLPARAMS** structure passed in to **lineMakeCall** via the **lpCallParams** parameter, so that if the line device uses a mechanism other than dialable digits to suppress call interrupts that that mechanism will be invoked.

**LINETRANSLATEOPTION_FORCELOCAL:** If the number is local but would have been translated as a long distance call (**LINETRANSLATERESULT_INTOLLLIST** bit set in the **LINETRANSLATEOUTPUT** structure), this option will force it to be translated as local. This is a temporary override of the toll list setting.

**LINETRANSLATEOPTION_FORCELD:** If the address could potentially have been a toll call, but would have been translated as a local call (**LINETRANSLATERESULT_NOTINTOLLLIST** bit set in the **LINETRANSLATEOUTPUT** structure), this option will force it to be translated as long distance. This is a temporary override of the toll list setting.

Finally the **lpTranslateOutput** parameters specifies a pointer to an application-allocated memory area to contain the output of the translation operation, of type **LINETRANSLATEOUTPUT.** Prior to calling **lineTranslateAdress,** the application should set the **dwTotalSizefield** of this structure to indicate the amount of memory available to TAPI for returning information.

## Return Values

Returns zero if the request is successful or a negative error number if an error has occurred. Possible return values are:

LINEERR_BADDEVICEID
LINEERR_INVALPOINTER
LINEERR_INCOMPATIBLEAPIVERSION
LINEERR_NODRIVER
LINEERR_INIFILECORRUPT
LINEERR_NOMEM
LINEERR_INVALADDRESS
LINEERR_OPERATIONFAILED
LINEERR_INVALAPPHANDLE
LINEERR_RESOURCEUNAVAIL
LINEERR_INVALCARD
LINEERR_STRUCTURETOOSMALL
LINEERR_INVALPARAM

## Placing a Call

You call this function to place a call on the specified line to the specified destination address. Optionally, call parameters can be specified if anything but default call setup parameters are requested.

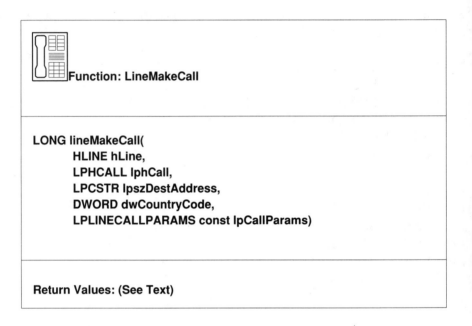

Function: LineMakeCall

```
LONG lineMakeCall(
    HLINE hLine,
    LPHCALL lphCall,
    LPCSTR lpszDestAddress,
    DWORD dwCountryCode,
    LPLINECALLPARAMS const lpCallParams)
```

Return Values: (See Text)

## Explanation of the Parameters and Usage

The hLine parameter specifies a handle to the open line device on which a call is to be originated.

The **lphCall** parameter specifies a pointer to an **HCALL** handle. The handle is only valid after the **LINE_REPLY** message is received by the application indicating that the **lineMakeCall** function successfully completed. Use this handle to identify the call when invoking other telephony operations on the call. The application will initially be the sole owner of this call. This handle is void if the function returns an error (synchronously or asynchronously by the reply message).

The **lpszDestAddress** handle specifies a pointer to the destination address. This follows the standard dialable number format. This pointer can be NULL for non-dialed addresses (as with a hot phone) or when all dialing will be performed using lineDial. In the latter case, **lineMakeCall** allocates an available call appearance that would typically remain in the dialtone state until dialing begins. Service providers that have inverse multiplexing capabilities may allow an application to specify multiple addresses at once.

The **dwCountryCode** specifies the country code of the called party. If a value of zero is specified, a default is used by the implementation.

The **lpCallParams** specifies a pointer to a **LINECALLPARAMS** structure. This structure allows the application to specify how it wants the call to be set up. If NULL is specified, a default 3.1kHz voice call is established and an arbitrary origination address on the line is selected. This structure allows the application to select elements such as the call's bearer mode, data rate, expected media mode, origination address, blocking of caller ID information, and dialing parameters.

## Return Values:

Returns a positive request ID if the function will be completed asynchronously, or a negative error number if an error has occurred. The dwParam2 parameter of the corresponding **LINE_REPLY** callback message is zero if the function is successful or it is a negative error number if an error has occurred. Possible return values are:

LINEERR_ADDRESSBLOCKED
LINEERR_INVALLINEHANDLE
LINEERR_BEARERMODEUNAVAIL
LINEERR_INVALLINESTATE
LINEERR_CALLUNAVAIL
LINEERR_INVALMEDIAMODE
LINEERR_DIALBILLING
LINEERR_INVALPARAM
LINEERR_DIALDIALTONE
LINEERR_INVALPOINTER
LINEERR_DIALPROMPT
LINEERR_INVALRATE
LINEERR_DIALQUIET
LINEERR_NOMEM
LINEERR_INUSE
LINEERR_OPERATIONFAILED
LINEERR_INVALADDRESS
LINEERR_OPERATIONUNAVAIL
LINEERR_INVALADDRESSID
LINEERR_RATEUNAVAIL

LINEERR_INVALADDRESSMODE
LINEERR_RESOURCEUNAVAIL
LINEERR_INVALBEARERMODE
LINEERR_STRUCTURETOOSMALL
LINEERR_INVALCALLPARAMS
LINEERR_UNINITIALIZED
LINEERR_INVALCOUNTRYCODE
LINEERR_USERUSERINFOTOOBIG

If LINEERR_INVALLINESTATE is returned, the line is currently not in a state in which this operation can be performed. A list of currently valid operations can be found in the **dwLineFeatures** field (of the type **LINEFEATURE_**) in the **LINEDEVSTATUS** structure. Calling **lineGetLineDevStatus** updates the information in **LINEDEVSTATUS**. If LINEERR_DIALBILLING, LINEERR_DIALQUIET, LINEERR_DIALDIALTONE, or LINEERR_DIALPROMPT is returned, none of the actions otherwise performed by **lineMakeCall** have occured; for example, none of the dialable addresses prior to the offending character has been dialed, no hookswitch state has changed, and so on.

After dialing has completed, several **LINE_CALLSTATE** messages are usually sent to the application to notify it about the progress of the call. No generally valid sequence of call-state transitions is specified, as no single fixed sequence of transitions can be guaranteed in practice. A typical sequence may cause a call to transition from dialtone, dialing, proceeding, ringback, to connected. With non-dialed lines, the call may typically transition directly to connected state.

An application has the option to specify an originating address on the specified line device. A service provider that models all stations on a switch as addresses on a single line device allows the application to originate calls from any of these stations using **lineMakeCall**.

The call parameters allow the application to make non-voice calls or request special call setup options that are not available by default. An application can partially dial using **lineMakeCall** and continue dialing using **lineDial**. To abandon a call attempt, use lineDrop. After **lineMakeCall** returns a success reply callback message to the application, a **LINE_CALLSTATE** message is sent to the application to indicate the current state of the call. This state will not necessarily be **LINECALLSTATE_DIALTONE**.

## Summary

The above essential functions provide an example of the programming model and philosophy of TAPI application building. The API contains dozens of functions which give the computer telephony programmer a wide field of operations and functionality.

My intention here was to provide you with a flavor of TAPI programming and introduce you to the functionality in the API. The objective was simple: "provide a snapshot of TAPI programming so that users can rapidly understand what it's all about."

You obviously need to be familiar, if not experienced, with the C and C++ languages to get the most out of the API. A word of advice: If you plan to do any serious TAPI programming the API will keep you very busy (you could, for example specialize in only certain regions of the API). No software engineer can expect to create anything sophisticated and full-featured without devoting time to a serious study of TAPI, eventually becoming a specialist at it. This can only be achieved with time and labor. As with any API, eventually you become part of it, knowing exactly how you want your applications to perform and interact with the API, the telephony environment, the operating system, and more. At that point you can graduate yourself as a TAPI programmer.

If, however, you are looking to create simple dialer or call handling applications, possibly for specific application in your enterprise, a high level telephony toolkit or collection of "plug-ins" may be more appropriate. The new breed of OLE (OCX and ActiveX) components may be what you need to speed, rapid application development and ease of use. Appendix B contains Object Pascal code as an example of what can be achieved with the Visual Voice components from Stylus Innovation.

# Appendix A

# The Official Windows Telephony Lexicon

## ADPCM

Adaptive Differential Pulse Code Modulation. ADPCM is a method of compressing data which calculates the difference between two consecutive speech samples in PCM-coded voice signals (see also PCM).

## Alerting (TAPI)

The Notifiction of the appearance of an inbound call by ringing on a POTS network or in a protocol message on an ISDN network.

## AGC

Automatic Gain Control. This is a voice processing algorithm that normalizes the volume level of recorded data to -6dBm. In telephony, it refers to the technique of automatically raising and lowering a transmitted signal to keep volume levels in a conversation even.

## Analog

Analog comes from the word analogous, which means "similar to." To understand the term better, think of an analog watch; the hands that turn clockwise represent, or are "similar to," the passing of time. The digital time-piece does not represent a comparison, a passing of time; it is a counter of seconds, minutes and hours.

## Analog Line

A standard phone line. Signals on an analog line use a set of standard in-band tones for call progress and DTMF signaling.

## Analog Transmission

Analog transmission means that the amplitude of the transmitted signal varies over a continuous range. Analog transmission (such as POTS signals) consists of sound traveling over lines as variations in an electrical current. It also refers to the transmission of analog signals without regard to the content. In other words, although the signal may be amplified, there is no immediate attempt to extract data from the signal.

Analog signals are very vulnerable to interference and noise on the line. They are also limited to the bandwidth of amplifiers, analog-to-digital converters, and other network equipment.

## ASI

Analog Station Interface. The circuitry that interfaces with an analog telephony station.

## Asynchronous Completion (Win32)

An application calls a function in the DLL, which then completes and returns to the application. If it completes immediately, this completion is known as synchronous, but if it is sent off to another system entity and the application goes on to other activities before the function completes (and the system later sends a message to the application announcing the function's completion), that completion is known as asynchronous.

If an application invokes multiple functions that complete asynchronously, their completion (reply) messages will not necessarily be received in the same sequence in which the functions were invoked. (This is the actual asynchronous nature of the function.)

## Asynchronous Transmission

Asynchronous Transmission is the process of transferring data at irregular intervals. Fax and email is transmitted asynchronously. The Internet is rapidly transforming from a purely asynchronous network in an Isochronous network as real-time Internet telephony takes shape.

## ATI

Analog Trunk Interface. The circuitry that interfaces with an analog telephone trunk.

## Audio Frequencies

This term refers to the range of frequencies that can be heard by the human ear.

## Audio Messaging Interchange Specification (AMIS)

AMIS is a set of standards that address the problem of how different voice messaging systems from various vendors can network or internetwork. AMIS allows these systems to share messages. An AMIS-compliant computer telephony system will thus allow you to install systems in a number of locations and conduct message interchange between the systems.

## Auto-attendant

There are several definitions of auto-attendant. In many parts of the world, it refers to the person who watches over a car at a nightclub, shopping complex, mall or public facility. It can also refer to the person who washes a car. In the U.S., this person is often known as the valet or car detailer if some washing and waxing is involved. The second definition refers to a computer telephony service or system (short for automated attendant) that stands in for a live telephonist.

I find the term a problem for me in Europe and elsewhere outside North America. In Israel, for example, I did some consulting for the King Solomon Hotel in Eilat. When I explained to the concierge (in my limited Hebrew) what I needed to look at in the PABX room, he brought me the valet. In the U.S., I find that many clients of mine also get confused with the term, and thus use "voice mail" as a generic term for all computer telephony systems and services, especially when referring to the automated attendant (and even if it doesn't take voice mail). I have thus switch to copious use of the term computerized telephonist.

## Automatic Speech Recognition (ASR)

Automatic speech recognition (ASR) technology reliably recognizes certain human speech, such as discrete numbers and short commands, or continuous strings of numbers, such as a spoken credit card number. Speaker-independent ASR can recognize a limited group of words (usually numbers and short commands) from any caller. Speaker-dependent ASR can identify a large vocabulary of commands from a specific speaker. Speaker-dependent ASR is popular in password-controlled systems and hands-free work environments (see also SIR and Voice Recognition below).

## BABT

British Approvals Board for Telecommunications. This is the British agency that approves telecommunications equipment.

## B channel

A 64 Kbps channel on an ISDN line that can carry voice or data (see also BRI).

## Bandwidth

The range of frequencies that a circuit can handle. With POTS, for example, the bandwidth is very narrow. The broader the range of frequencies, the more information the line can handle. The typical POTS circuit has a bandwidth of 3,100 Hz centered between 300 Hz and 3,400 Hz.

## Bearer Mode

The type of coding, or compression, which the telephone network is permitted to perform on the bit stream carried on the bearer channel. In POTS, the bearer mode will always be 3.1 kHz voice. The "speech" bearer mode is the most compressible, "voice" less so, and so on. A data bearer mode implies that the data stream will not be compressed by the network (the connection is "clear channel").

## Bearer Services

Services designed to transfer information from point A to point B.

## BRI-ISDN

A CCITT-defined "Basic-Rate Interface" ISDN connection consisting of two B channels of 64 Kbps each for voice or data, and one D channel of 16 Kbps for control (2B+D).

## Busy

Busy means that a telephone service is in use. You get two busies in telephony. Normal busy refers to the off-hook state of a telephone, which signifies that a connection has been set up between subscriber and the CO. Fast busy refers to the "blocked" state of a network (see the Erlang B entry below).

## Cadence

A repeating cycle. In computer telephony, it refers to the cycle of tones and silence intervals in the audio signal. In the U.S., a ringing tone cadence is typically one second of tone followed by three seconds of silence.

## Call

Two or more parties exchanging information (or attempting to exchange information) using telephony equipment. Many of the functions in TAPI and in the Telephony SPI operate on calls.

## Call Progress

Setting up a telephone call goes through several phases. Taking the phone offhook returns dial tone to indicate that a number can be dialed. Hearing the dial tone, the user dials the desired number. When the call reaches the destination phone, the caller will either receive a busy indication, indicating the called number is busy, or a ring back indication, indicating the dialed party is being alerted. Call progress is the process of monitoring the progress of a call through the various stages.

In analog networks, audible tones generated by the network provide the call progress indications to the user. Different tones allow the human ear to interpret the progress of the call, and telephony hardware for use on POTS networks is often designed to detect these tones as well. On digital networks (such as PBX or ISDN), the network may send indication messages (normally through a D-channel protocol) to the phone to indicate the status of the call, and the phone may generate most tones locally, driven by those messages.

## Callback Function

A callback is a mechanism through which an application is notified of events outside the application. The callback function is a function in the application that is called by Windows to inform the application that a message has arrived for it.

NOTE: A developer who creates a function can choose to make it a callback function. Functions defined in an API have already been defined to be (or not be) callback functions, and must be used accordingly.

## Called-ID

An identification (number, name) of the party being called. This identification is of interest when you transfer or forward a call. For example, when an unanswered call is forwarded to a voice messaging system, the called-ID of the original call is used to locate the mailbox of the called party.

## Caller-ID

An identification (number, name) of the party initiating a call, as displayed to the called party prior to answering the call. A caller-ID may also be either unknown (due to telephone switch limitations), blocked (concealed by the caller), or not yet known, but received later.

## Calling Card (TAPI)

To Telephony, a calling card is not just a means for billing calls, it represents a distinct dialing procedure. For example, an applications can use calling-card functions to present a menu of pre-configured calling cards to the user. The user's response determines the proper dialing sequence for the call, which tells TAPI whether to make the call with the default carrier, override the default carrier and use a given calling code, or use another specific dialing sequence.

## Call Progress Tone

The call progress tone is an audible tone that indicates the progress of a telephone call. Call progress tones include busy tones and ring tone. (See Chapter 1 and Chapter 2.)

## CCITT

*Cimite' Consultatif Internationale de Telegraphique et Telephonique*, or the International Telephone and Telegraph Consultative Committee. This committee sets the international standards for data communications.

## Central Office

A public switch, directly connected to a number of telephones in a given geographical area.

## Centrex™

A service provided by central offices that provides a virtual PBX to a set of extensions. It offers features such as transfer, conference, and forward within that set of extensions.

## CEPT

Conference on European Posts & Telecommunications. This committee sets worldwide standards for data communications.

## Channel

In computer telephony (voice processing), a channel is a logical path that is used for voice processing (see also entries for Port and Line to avoid confusion).

## Client Computer (TAPI)

The host environment for which the Windows Telephony API is defined. This is the computer running the Windows operating system version 3.1 or higher.

## CO

CO stands for central office. This is the telephone company location that houses switching gear and other services for local telephone subscribers. Outside North America, the equivalent term is telephone exchange (or local telephone exchange).

## CO Lines

CO Lines are the telephone lines that connect CPE to the telephone company.

## Codec

The codec is a device that converts an analog signal into a digital signal and vice versa.

## Compression

Compression means reducing the representation of information but not the information itself. Voice data that is recorded and stored at high resolution is referred to as linear and is thus uncompressed. Compression saves transmission time and storage space. For voice processing purposes, data has to be compressed to PCM or ADPCM (see also Linear Data below).

## Conference

The ability to join three or more people in a telephone call. Certain privileges may apply to a conference call. For example, callers may be allowed to listen to the conference but not speak.

## CPE

Customer-Premises Equipment or Customer Premise Equipment. CPE is any piece of telecommunications equipment that is not owned by the CO. A PBX or ACD unit is CPE. The telephone set or telephone devices are also CPE.

## CSID

CSID means customer subscriber identification — the remote devices telephone number.

## D Channel

A Channel on an ISDN line that can carry signaling information and low-speed packet data.

## Desktop (TAPI)

The logical pairing of a user's computer and telephone.

## Device (TAPI)

The device is a hardware or logical entity referred to in software engineering or computer telephony language. DOS and Windows view each board as a device, an entity, that can be attached to the computer.

## Device Driver

The device driver is a collection of subroutines and data that provides an interface between the operating system and the application that needs to make use of the device, or drive it.

## Dial Tone

Dial tone is the sound you hear upon lifting the receiver (not when you lift the telephone). The signal is usually transmitted in the frequency range 300Hz to 450Hz. It signifies the "all clear" to dial a number (see Chapter 1).

## DID

DID is the ability to directly dial into a company extension without going through a computer telephonist (automated attendant) or a live telephonist.

## Digital Line

A digital station line on a PBX or digital-key system. Signaling on a digital line uses vendor-specific (proprietary) protocols to exchange messages between the switch and the phone. A digital line typically requires a "matched" phone set.

## Digital Signal Processing (DSP)

For the past ten years or more voice processing components have made it possible for humans to interact and have a dialog (in an almost natural conversation) with computers. This dialog has taken place over the telephone and "face to face" with the computer the key input device being a microphone. The most natural form of communication between humans is voice. If humans and computers are to co-exist (now don't laugh) it seems fitting that humans converse with computers in the style to which humans are accustomed (just like you see on Star Trek, Commander Data is walking collection of DSP components).

This is where the DSP chip steps in. It tries to make sense of analog-turned-digital information that it collects from the surrounding world. Like the processes of the human brain it then communicates this info, as binary code, to the computer for response (the processing is very similar to the way living beings use their senses to process information). The following is a rudimentary description of DSP in action.

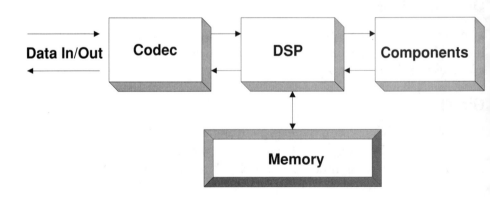

**Figure A1:** When an analog signal arrives at the voice processing port (the telephone connection on the voice processing card), the port feeds it to a coding/decoding device (the codec). The codec changes the analog wave form into digital data (a collection of 1s and 0s), which is the language that computers speak and understand. By being able to perform mathematical calculations at incredible speeds (billions of calculations per second), the DSP chip is able digitize and compress analog audio into digital information.

Digital signal processing technology is what lies at the heart of all information processing and computer telephony. After the analog information comes out digital from the codec, the DSP chip analyzes this digital data and begins to perform calculations to make sense of what it represents. The DSP chip processes the signal by blazing through billions of mathematical operations per second. This processing is controlled by the DSP operating system, or DSP kernel, that loads onto the telephony card or DSP card at runtime. (That the DSP chip has its own operating system is another great attribute of this marvelous invention, however it takes the best engineers to write the code that run these beauties.)

Just recognizing that a human, rather than an answering machine, said "hello," involves countless calculations. In computer telephony circles this is known as positive voice detection or control (see the entry for PVC below) If you had to sit down with a fancy scientific calculator and try and duplicate the effort manually, it would take you years to finish. But the DSP does this task in a sliver of time (it gives the software something to match against databases or repositories of information). Once the DSP finishes processing the signal, the computer telephony software picks up the result and determines whether it should say "hello" back or hang up. The technology (voice detection that is) is valuable for telemarketing. It allows telemarketing dialers to abandon the call when an answering machine on the other end of the line picks up the call.

The processing power of these components is astonishing: currently, they can handle approximately two billion operations per second (2 BOPS); by the end of the century, you may be plugging in devices that can process 4 BOPS. Already, interactions between computers and humans are becoming so lucid that it's sometimes hard to tell whether you're talking to a machine or a human.

One of the leading manufacturers of DSP chips is Texas Instruments. You can find its technological touch on all leading computer telephony components. One of this company's top products is a component called the MVP. MVP stands for Multimedia Video Processor. The chip is better known in DSP programming and computer telephony circles as the TMS320C80. I won't go into the technical underpinnings of this chip because that would add a few thousand pages to the appendix, except to say that it can process 2 BOPS and run cool without constantly needing a fan on its back, unlike some of the high-end microprocessors on PC motherboards, like the Pentiums. One of the important applications that this chip will enable is real-time video over wide area communications, such as the Internet.

Thus, the DSP chip has been the chief enabling technology in the voice processing industry. By using this chip, we can stick all this techno-wizardry onto cards that plug into PCs. You could say that the DSP chip has been the driving force behind computer telephony.

## Digitization

Digitization is the process of converting analog signals into digital representations of that signal.

## Digits

Telephony digits:

> 0 through 9
> A though D
> * (star)
> # ("pound" also known as "hash")

## DTMF

DTMF means dual-tone multifrequency. It refers to the tone used in dialing (also known as "touch tone"). When you depress the key or button on the telephony set or click on the button on your "soft-dialer," you are not transmitting one tone but two. This combination of tones is where the term dual comes in. Figure A-2 illustrates the frequencies used on the world's touch-tone telephones.

In analog networks, audible tones generated by the network provide the call progress indications to the user. Different tones allow the human ear to interpret the progress of the call. On digital networks (such as PBX or ISDN), the network may send indication messages to the phone to indicate the status of the call, and the phone may generate most tones locally, driven by those messages.

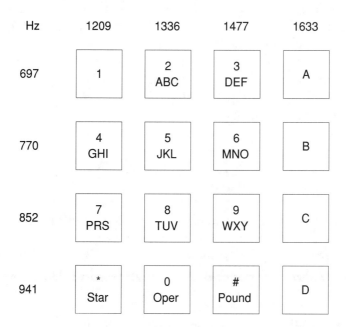

**Figure A-2:** The DTMF frequencies used on the world's touch-tone telephones. The A,B,C, and D tones are not usually found on common telephone sets.

## DTMF Cut-Through

Cut-through refers to the ability of voice processing systems to detect and act on touch tones during the playing of a voice. The tone can either suspend the playing of a file or it can be used to replay or forward to portions of it. Not only can the digit be detected but the information can be passed to software control in order to execute an a new event, such as transferring to a telephonist.

## ECTF

The Enterprise Computer Telephony Forum (ECTF) is an industry organization formed to foster an open, competitive market for Computer Telephony technology. Participants include industry suppliers, developers, system integrators, and users working to achieve agreement on multi-vendor implementations of computer telephony technology based on international defacto and dejour standards.

Principal Members of the ECTF are: Amtelco, Amteva, Inc., Apple Computers Aspect Telecommunications, AT&T, Bellcore, Brooktrout

Technology, Centigram Communications, Cintech Tele-Management, CSELT, Database Network Services, Dialogic Corporation, Dianatel Corporation, Digital Equipment Corporation, Ericcson Business Networks AB, Fujitsu Limited, Hewlett-Packard Company, IBM Corporation, InterVoice, Mitel Corporation, Motorola, Inc., Natural MicroSystems Corp., Networks Unlimited AG, Nortel, Novell, Rockwell Telecommunications, Siemens AG, Sun Microsystems, Unimax, and Voicetek Corporation.

The major goal of the ECTF S.100 Interoperability is to provide users with more choices for CTI-enabled applications.

In March 8, 1996 the ECTF announced the first in a series of computer-telephony integration (CTI) technical implementation agreements. The ECTF Computer Telephony Services Framework, known as ECTF S.100 (rev 1.0 Media Services APIs), will accelerate the deployment of CTI technology by facilitating interoperability between and among different suppliers' equipment.

The ECTF S.100 APIs are derived from the SCSA TAO framework (see SCSA and SCSA TAO below), the Distributed Computer Telephony (DCT) architecture, and many other valuable contributions from multiple members.

**Open Environment is Key to Interoperability**

The ECTF S.100 agreement provides an effective way to develop computer telephony applications in an open environment. The agreement defines a client-server model in which client applications use a collection of services to allocate, configure, and operate hardware resources. The agreement separates implementation details of call processing and switch hardware to enable portable applications to be written. It furnishes these services via operating system-independent application program interfaces (APIs) that may be extended to support customized APIs.

The S.100 API Definition is offered as a Trial Use Implementation Agreement. Trial Use documents are a step in the development of a permanent interoperability agreement. The ECTF will review this document periodically and issue revisions.

**The Target CT Environment**

The ECTF has determined that by defining a CT framework with a few general-purpose layers with carefully specified interfaces, the market complexity of incompatible CT subsystems can be simplified. In addition, if these few components can efficiently deal with the wide range of real-time, asynchronous, distributed CT applications, the technical complexity of integrating broad-based CT applications can be significantly reduced.

The goals of the ECTF are to provide a framework that allows Client/Server or Server-Resident CT Implementations: Allows application developers to target their applications to reside either on a single server that contains the CT hardware, or in a distributed client-server solution.

Application Portability: Provides application developers the confidence that applications will operate in a standard, predictable way across multiple client or server environments. Application developers should expect that a compliant application will operate with any compliant server equipped with the appropriate features required by the application.

Application Modularity: Multiple CT applications should be able to operate fully in the framework without knowledge of each other. This includes cooperating on calls and sharing resources.

Multiple Applications Sharing a Call: Multiple applications should cooperate and share a single call, each one in sequence. Each compliant application, when finished with a call, should pass the call in a cooperative way to the next application.

## EIN

Enterprise information network. Many IT specialist refer to the "enterprise network," which refers to the LAN. In recent years, the flow of enterprise information now traverses interconnecting corporate networks, wide area networks, secure data links between offices and associated companies, public networks (such the Internet), and the telephone network. I thus use the term EIN to refer to all telecommunications networks used by the enterprise to manage and provide access to information.

## Erlang B Carried-Traffic Table

This is a table used to determine port requirements for telephony and computer telephony services. You need to first determine your Erlang value for the specific group of telephone lines coming into your service. A Ports Requirements software suite, which automates the establishment of values through simulation of traffic and usage patterns and automatically calculates port requirements, is available from the author at no charge (see Introduction). See Figure A-3 for the table that contains these Erlang values.

The following text was reprinted with Permission by IDG Books Worldwide Inc., from the book Computer Telephony Strategies (Shapiro, Jeffrey R.., Computer Telephony Strategies, IDG Books Worldwide Inc., 1996.)

An Erlang is a measurement of telephone traffic in hours, and was named after Danish telephone engineer A. K. Erlang, whose forte—you guessed it—was call-traffic engineering. Why do you have to find your Erlangs? Well, one of the goals of this section is to help solve one of the biggest puzzles that any fledgling CT project or call center campaign has—determining the number of ports you will need.

But why do you need to determine the number of ports? Why should you even care? The reason is that this information can help you when performing the following tasks:

   Establishing needs
   Determining number of ports for the service
   Determining how much the system will cost
   Justifying the cost
   Predicting amortization and establishing a return on investment
   Preparing equipment purchases
   Formulating budget and ordering necessary hardware and software
   Extending or expanding existing systems (such as the PBX)
   Ordering more telephone lines

These factors are all important decisions that you have to make when deployment and purchasing process of computer telephony. To make the correct decisions, you need data, and you need data that you can trust.

The process to get that data is not unlike the effort that road traffic engineers go through to determine how much traffic a stretch of highway can handle, or what its peak hour is. The road traffic engineers lay a cable across the road, and every time that a vehicle drives over it, a counter records the event. The engineers can then compare the data acquired on one road to the data acquired from other routes and than plan roadway needs accordingly.

In computer telephony, we use various tools to do the same thing with call traffic. We also try to predict calling patterns, which is a lot tougher than counting calls. This excerpt tells you how to set out to acquire this data and how to use the data when developing or buying and deploying CT systems.

Before discussing the tasks that you need this data for, as listed earlier, we'll get into some background information, do a quick exercise to determine some Erlang values, and, thereby, determine the number of ports needed for your computer telephony services. Before you can ask me for my definition of a port, here it is (as it relates to a computer telephony system and this exercise).

Many telephony or telecommunications gurus argue about the proper definitions for channels, ports, lines, trunks, conferencing, and so on, but within the scope of this book, you can basically view a port as the number of "electronic doorways" that you need to make your enterprise or service accessible to a caller. The ports on your PBX or switching service are the entrances they have via the telco lines or telephone extensions. Ports also include the telephone entrances to the voice processing system, the fax server, the call sequencer, and the ACD unit. Other ports include the entrances to your popular Web server that 100 million Netizens are trying to "hit."

Although you have ports on all the telephony devices in the enterprise, the number of callers who can connect to your service or services depends on the number of lines that enter your facility. For example, you may have 100 telephone lines entering your PBX and several hundred extensions, of which you've dedicated 24 to the voice processing system.

But let's get back to call-traffic engineering: I learned an Erlang forecasting technique for determining Erlang values from the telephony and PBX community and then successfully used it for computer telephony. This procedure is often used to forecast the number of PBX ports and lines needed by the enterprise, based on call information provided by the central office or telephone exchange.

We'll get into the specifics of this forecasting method later, but here's how it basically works: The telephone company monitors call traffic on your group of telephone lines and determines an Erlang value for the group. Your call traffic is measured in Erlangs per hour. In other words, if your group of 20 lines carries 15.55 Erlangs per hour, you're using just over 15 of your lines in your peak calling hour.

This type of call-traffic engineering is as critical for telemarketing campaigns and call centers as it is for the telephone company. New technologies are making call-traffic engineering a little easier to perform. One school of thought is that computer simulation should be applied to telecommunications traffic management. Computer simulation is very expensive, however: One of the best simulation suites you can get costs a fortune, and the actual simulation would cost more than the system it recommends.

Many believe that applying logistics—doing the math; using common sense, sixth sense, and reason; and drawing on experience—still works best. Many consultants try to monitor lines using an assortment of electronic equipment, such as current detection devices; but this method is just not practical for forecasting call traffic for a large installation, in which call patterns may widely vary.

Determining the number of ports you will need is an important task you must apply to all of the CT service domains. As such, although the consultant or vendor can help you determine your needs, you, the IT executive or enterprise CT engineer, should be familiar with the analysis and needs procedure in order to plan accordingly.

By first understanding what Erlang set out to accomplish, you will discover why this section is one the most important technical considerations for any computer telephony or Windows telephony project. At the dawn of the Twentieth Century, when telephone traffic started to go ballistic, Erlang used his engineering skills to determine how to deploy telephone exchanges. The dilemma he set his sights on was figuring out a way to predict the hardware necessary to handle the widely fluctuating call traffic; if not enough ports were available to switch lines, callers would be blocked from obtaining service.

Blocking is another telephony term you need to know about (from the deployment angle), as it applies to several concepts. A call is blocked when the caller gets busy tone—or, as they say in Europe, an engaged signal. Essentially, the service you were trying to reach is unavailable because someone else got there before you—or so you think. What you don't know is that back at the telephone company, the engineers are getting ulcers because their subscribers, who expect to hear a dial tone when they lift their handsets or press 9 to get service from the PBX, cannot be serviced because all possible call paths to connect them are in use.

In small telephone exchanges, it's not uncommon for subscribers to receive an exchange busy" signal. Even the big exchanges can max out. Remember the San Francisco and Los Angeles earthquakes? California calls were blocked for many hours.

The concept behind the PBX catering to 100 extensions and a telephone exchange catering to tens of thousands of subscribers works the same. Computer telephony software also has to take call blocking into consideration. A system trying to seize a line to deliver a message or make a call may be blocked because every outbound telephone line is in use. The CT software has to cater to two scenarios: The system was often blocked because all internal lines were in use (indicated by service unavailable or reorder tone 1) or because the exchange itself was unable to provide the PBX with dial tone (indicated by service unavailable or reorder tone 2).

Call blocking also refers to the incapability to make a call (which has nothing to do with a busy service). For example: A person who tries to make a long-distance call from the enterprise or a hotel may be blocked from incurring long distance charges. The term blocked also refers to the caller who receives a polite "take a hike" message from someone who uses Caller ID. (By the way, Caller ID is great for blocking unwanted calls from in-laws.)

Finally, the term applies to digital PBX systems. A non-blocking PBX refers to the routing technology used to move voice and data through a digital PBX.

You can view the exercise that you'll perform later from another angle: how much you need to spend. Ports usually come on interface cards to the PBX or CT system, and nowadays, you pay by the card. The more ports you have, the less risk you run of blocking callers and ruining that alluring image you want to protect, especially if you provide the computer telephonist, automated attendant services, or voice mail facilities.

For example, I recently called on a large public company in New Jersey that was having a problem with its voice mail system. They use a few human telephonists to route calls manually, and the calls then divert to voice mail if the party is unavailable. As it turns out, I didn't have to visit the site to determine the problem. The telephonist told me that the system works at certain times of the day, but not during the peak calling periods.

"The voice mail refuses to take the calls, including the ones I transfer directly, and then they return to the console," she said. "We have recommended to the CIO that he throw out the system because there is obviously a bug in the software."

The telephonist and her coworkers were wrong in their assessments, however; what was really happening is that the callers were being blocked from leaving messages at busy times because the firm did not have enough voice mail ports to handle the voice mail traffic. The PBX engineer correctly catered to this possibility by diverting the unserviced subscribers back to the telephonist consoles, rather than letting the calls ring endlessly at the voice mail ports or get busy signals. There was nothing wrong with the software;

the firm just had not anticipated or planned for the increase in traffic.

Be that as it may, you do not want to invest in equipment that "maxes out" for only a short while and then stands idle for the rest of the day. This situation is exactly the type of problem that you can avoid by applying Erlang's theories.

The following exercise will help you understand your telephone line requirements when dealing with the telephone companies. This list represents the steps that you take in the port forecasting process:

1. Forecast the number of calls in the peak hour
2. Determine the average call length
3. Determine the Erlang Value
4. Determine the blocking tolerance level ( the acceptable pecentage of people who won't be able to connect to your service)
5. Determine the number of ports

Let's perform a quick exercise in port forecasting so that you get the idea. (Then we'll discuss some factors that illustrate that this science is not an exact, but rather an exercise in educated guessing.)

Consider the math: in telephone traffic engineering, one Erlang, the known value, is equal to one hour of telephone conversation. Multiply 60 minutes by 60 seconds to get a value that you can apply to an equation.

1 Erlang = 3,600 seconds of telephone conversation.

Thus, if we can establish the number of calls on a network and multiply this by the average length of a call, and then divide this product by one Erlang, we arrive at the number of Erlangs carried by the network in an hour. The equation looks like this:

$(a * b) / 3{,}600 = x$

Where a = number of calls, b = average length of call (in seconds), and x = number of Erlangs

Let's say that we have a trunk group (a collection of telephone lines) and have established that the number of peak hour calls to the service is 1,000. After some further analysis, we also determine that the average call time is 120 seconds (two minutes). Now back to the equation:

$(1{,}000 * 120) / 3{,}600 = 35.27$ Erlangs

Thus, our group of lines is carrying 35.27 Erlangs. Translated into lay terms, this means that for the busiest hour, 35 lines in the trunk group were carrying conversations.

Does that mean that we now have to have no less than 35 ports (or at least 50) to make sure that every caller is serviced by the equipment? No, we are not that far yet. We have to take blocking into account. This part of the forecasting method is known as Erlang B theory.

Erlang B is the probability theory developed by Erlang to estimate the number of lines needed to carry telephone traffic, going on the assumption that callers will abandon the call if they are not serviced quickly. (There are many other factors to consider, especially if you bring sequencing and hold facilities into the picture. If you want to study this further, there are a number of books you can consult. The bibliography at the back of this book is a good place to start.)

No two enterprises are the same; as such, no two call processing campaigns or computer telephony projects are the same. Just how many blocked or unserviced calls you feel is tolerable depends on the situation and the application—and the business plan.

For example, If you are planning a sales campaign that incorporates some very costly publicity, you may feel 1 blocked call in a thousand is the limit, or .1 percent. Alternatively, the sales room manager would freak out if one out of every ten calls to the enterprise is blocked, because he or she might have sales agents waiting for calls and no one can get through. (One Connecticut financial services firm hired a consultant to find out why its blocked-call levels were going through the roof. It was later discovered that the IT department gave staff access to the World Wide Web and someone had found a link to a very graphic home page at Amsterdam's Red Light district.) However, if your application is not as critical as a sales campaign, then one blocked call in ten, or 10 percent, may be acceptable.

With the call-blocking tolerance in percentage terms in hand, and armed with the Erlang value we calculated in the preceding exercise, we need to reference the Erlang B Carried-Traffic Table in Figure A-3.

# Figure A-3: The Erlang B carried-traffic table.

| Ports | 10% | 5% | 2% | 1% | .5% | .1% | Ports | 10% | 5% | 2% | 1% | .5% | .1% |
|---|---|---|---|---|---|---|---|---|---|---|---|---|---|
| 4 | 2.05 | 1.52 | 1.09 | 0.87 | 0.70 | 0.44 | 29 | 27.05 | 23.83 | 21.04 | 19.49 | 18.22 | 15.93 |
| 5 | 2.88 | 2.22 | 1.66 | 1.36 | 1.13 | 0.76 | 30 | 28.11 | 24.80 | 21.93 | 20.34 | 19.04 | 16.58 |
| 6 | 3.76 | 2.96 | 2.28 | 1.91 | 1.62 | 1.15 | 31 | 29.17 | 25.77 | 22.83 | 21.19 | 19.86 | 17.44 |
| 7 | 4.67 | 3.74 | 2.94 | 2.50 | 2.16 | 1.58 | 32 | 30.23 | 26.75 | 23.73 | 22.05 | 20.68 | 18.20 |
| 8 | 5.60 | 4.54 | 3.63 | 3.13 | 2.73 | 2.05 | 33 | 31.30 | 27.72 | 24.63 | 22.91 | 21.51 | 18.97 |
| 9 | 6.55 | 5.37 | 4.34 | 3.78 | 3.33 | 2.56 | 34 | 32.36 | 28.70 | 25.53 | 23.77 | 22.34 | 19.74 |
| 10 | 7.51 | 6.22 | 5.08 | 4.46 | 3.96 | 3.09 | 35 | 33.43 | 29.68 | 26.43 | 24.64 | 23.17 | 20.52 |
| 11 | 8.49 | 7.08 | 5.84 | 5.16 | 4.61 | 3.65 | 36 | 34.50 | 30.66 | 27.34 | 25.51 | 24.01 | 21.30 |
| 12 | 9.47 | 7.95 | 6.62 | 5.88 | 5.28 | 4.23 | 37 | 35.57 | 31.64 | 28.25 | 26.38 | 24.85 | 22.09 |
| 13 | 10.47 | 8.83 | 7.41 | 6.61 | 5.96 | 4.83 | 38 | 36.64 | 32.63 | 29.17 | 27.25 | 25.69 | 22.86 |
| 14 | 11.47 | 9.73 | 8.20 | 7.35 | 6.66 | 5.45 | 39 | 37.71 | 33.61 | 33.08 | 28.13 | 26.54 | 23.65 |
| 15 | 12.48 | 10.63 | 9.01 | 8.11 | 7.38 | 6.08 | 40 | 38.79 | 34.60 | 31.00 | 29.01 | 27.38 | 24.44 |
| 16 | 13.50 | 11.54 | 9.83 | 8.87 | 8.10 | 6.72 | 41 | 39.86 | 35.59 | 31.92 | 28.89 | 28.23 | 25.24 |
| 17 | 14.52 | 12.46 | 10.66 | 9.65 | 8.83 | 7.38 | 42 | 40.94 | 36.58 | 32.84 | 30.77 | 29.08 | 26.04 |
| 18 | 15.55 | 13.38 | 11.49 | 10.44 | 9.58 | 8.05 | 43 | 42.01 | 37.57 | 33.76 | 31.66 | 29.94 | 26.84 |
| 19 | 16.58 | 14.31 | 12.33 | 11.23 | 10.33 | 8.72 | 44 | 43.09 | 38.56 | 34.58 | 32.54 | 30.80 | 27.64 |
| 20 | 17.61 | 15.25 | 13.18 | 12.03 | 11.09 | 9.41 | 45 | 44.16 | 39.55 | 35.61 | 33.43 | 31.66 | 28.45 |
| 21 | 18.65 | 16.19 | 14.04 | 12.84 | 11.86 | 10.11 | 46 | 45.24 | 40.54 | 36.53 | 34.32 | 32.52 | 29.96 |
| 22 | 19.69 | 17.13 | 14.90 | 13.65 | 12.64 | 10.81 | 47 | 46.32 | 41.54 | 37.46 | 35.21 | 33.38 | 30.07 |
| 23 | 20.74 | 18.08 | 15.76 | 14.47 | 13.42 | 11.52 | 48 | 47.40 | 42.54 | 38.39 | 36.11 | 34.25 | 30.88 |
| 24 | 21.87 | 19.03 | 16.63 | 15.29 | 14.20 | 12.24 | 49 | 48.48 | 43.54 | 39.32 | 37.00 | 35.11 | 31.69 |
| 25 | 22.83 | 19.99 | 17.50 | 16.12 | 15.00 | 12.97 | 50 | 49.56 | 44.53 | 40.25 | 37.90 | 35.98 | 32.51 |
| 26 | 23.88 | 20.94 | 18.38 | 16.96 | 15.80 | 13.70 | 51 | 50.60 | 45.50 | 41.20 | 38.80 | 36.85 | 33.30 |
| 27 | 24.94 | 21.90 | 19.26 | 17.80 | 16.60 | 14.44 | 52 | 51.70 | 46.50 | 42.10 | 39.70 | 37.72 | 34.20 |
| 28 | 26.00 | 22.87 | 20.15 | 18.64 | 17.41 | 15.18 | 53 | 52.80 | 47.50 | 43.10 | 40.60 | 38.60 | 35.00 |

| Ports | 10% | 5% | 2% | 1% | .5% | .1% | Ports | 10% | 5% | 2% | 1% | .5% | .1% |
|---|---|---|---|---|---|---|---|---|---|---|---|---|---|
| 54 | 53.90 | 48.50 | 44.00 | 41.50 | 39.47 | 35.60 | 79 | 81.10 | 73.80 | 67.70 | 64.40 | 61.77 | 57.00 |
| 55 | 55.00 | 49.50 | 44.90 | 42.40 | 40.35 | 36.60 | 80 | 82.20 | 74.80 | 68.70 | 65.40 | 62.67 | 57.80 |
| 56 | 56.10 | 50.50 | 45.50 | 43.50 | 41.23 | 37.50 | 81 | 83.30 | 75.80 | 69.60 | 66.30 | 63.57 | 58.70 |
| 57 | 57.10 | 51.50 | 46.80 | 44.20 | 42.11 | 38.80 | 82 | 84.40 | 76.90 | 70.60 | 67.20 | 64.49 | 59.50 |
| 58 | 58.20 | 52.60 | 47.80 | 45.10 | 42.99 | 39.10 | 83 | 85.50 | 77.90 | 71.60 | 68.20 | 65.39 | 60.40 |
| 59 | 59.30 | 53.60 | 48.70 | 46.00 | 43.88 | 40.00 | 84 | 86.60 | 78.90 | 72.50 | 69.10 | 66.29 | 61.30 |
| 60 | 60.40 | 54.60 | 49.60 | 46.90 | 44.76 | 40.80 | 85 | 87.70 | 79.90 | 73.50 | 70.00 | 67.20 | 62.10 |
| 61 | 61.50 | 55.60 | 50.60 | 47.90 | 45.64 | 41.60 | 86 | 88.80 | 80.90 | 74.50 | 70.90 | 68.11 | 63.00 |
| 62 | 62.60 | 56.60 | 51.50 | 48.80 | 46.53 | 42.50 | 87 | 89.90 | 82.00 | 75.40 | 71.90 | 69.02 | 63.90 |
| 63 | 63.70 | 57.60 | 52.50 | 49.70 | 47.42 | 43.30 | 88 | 91.00 | 83.00 | 76.40 | 72.80 | 69.93 | 64.70 |
| 64 | 64.80 | 58.60 | 53.40 | 50.60 | 48.31 | 44.20 | 89 | 92.10 | 84.00 | 77.30 | 73.70 | 70.85 | 65.60 |
| 65 | 65.80 | 59.60 | 54.40 | 51.50 | 49.19 | 45.00 | 90 | 93.10 | 85.00 | 78.30 | 74.70 | 71.75 | 66.50 |
| 66 | 66.90 | 60.60 | 55.30 | 52.40 | 50.09 | 45.80 | 91 | 94.20 | 86.00 | 79.30 | 75.60 | 72.67 | 67.40 |
| 67 | 68.00 | 61.60 | 56.30 | 53.30 | 50.98 | 46.70 | 92 | 95.30 | 87.10 | 80.20 | 76.20 | 73.58 | 68.20 |
| 68 | 69.10 | 62.60 | 57.20 | 54.30 | 51.87 | 47.50 | 93 | 96.40 | 88.10 | 81.20 | 77.50 | 74.49 | 69.10 |
| 69 | 70.20 | 63.70 | 58.20 | 55.20 | 52.77 | 48.40 | 94 | 97.50 | 89.10 | 82.20 | 78.40 | 75.41 | 70.00 |
| 70 | 71.30 | 64.70 | 59.10 | 56.10 | 53.66 | 49.20 | 95 | 98.60 | 90.10 | 83.10 | 79.40 | 76.33 | 70.90 |
| 71 | 72.40 | 65.70 | 60.10 | 57.00 | 54.56 | 50.10 | 96 | 99.70 | 91.90 | 84.10 | 80.30 | 77.24 | 71.70 |
| 72 | 73.50 | 66.70 | 61.00 | 58.00 | 55.46 | 50.90 | 97 | 100.8 | 92.90 | 85.10 | 81.20 | 78.16 | 72.60 |
| 73 | 74.60 | 67.70 | 62.00 | 58.90 | 56.35 | 51.80 | 98 | 101.9 | 93.20 | 86.00 | 82.20 | 79.07 | 73.50 |
| 74 | 75.60 | 68.70 | 62.90 | 59.80 | 57.25 | 52.70 | 99 | 103.0 | 94.20 | 87.00 | 83.10 | 79.99 | 74.40 |
| 75 | 76.70 | 69.70 | 63.90 | 60.70 | 58.15 | 53.50 | 100 | 104.1 | 95.20 | 88.00 | 84.10 | 80.91 | 75.20 |
| 76 | 77.80 | 70.80 | 64.90 | 61.70 | 59.05 | 54.40 | | | | | | | |
| 77 | 78.90 | 71.80 | 65.80 | 62.60 | 59.96 | 55.20 | | | | | | | |
| 78 | 80.00 | 72.80 | 66.80 | 63.50 | 60.86 | 56.10 | | | | | | | |

From the table, you can determine that a blocking tolerance of one percent requires at least 47 telephone lines for an Erlang value of 35.27, and at least 37 ports for a tolerance of ten percent. A campaign manager in a call center might opt to install 45 ports just to be "on the safe side." Keep in mind, however, that being on the safe side can be costly.

Now let's discuss some variables:

**Forecasting the number of calls in the peak hour.**

This task is no easy matter, especially when you are not practiced in it, or are starting a new campaign for which you have very little data, or are setting up a new voice processing or voice response system or designing a telephony system from scratch. Enterprises throughout the world engage the services of their telephone companies, which can provide such data. The telco can keep tabs on the inbound call traffic and blocking levels on your group of lines.

Often, however, this service does not come cheap, and many telcos do not have the staff or equipment to provide you with more advanced data such as peak hours and calling patterns. Some companies may also refuse such requests from small businesses. In many parts of the world, this data is unavailable, period. The last problem was the case in South Africa before the telephone department was privatized.

In telemarketing campaigns or call center applications, the number of calls per hour varies according to factors such as extent of paid-for publicity, media coverage, nature of the product being offered, application, and so on. In voice processing or voice response applications, forecasting the number of calls per hour is even more difficult because there are so many unknown factors that influence the calling patterns.

Trying to pinpoint the peak hour so that you can count the calls manually is no cinch, either—believe me, I know. I have analyzed informal call center sites where the hour after the doors open in the morning is the busiest period. Other organizations experience their busiest hour after 2 p.m. International or national call centers can experience their busiest hour as late as midnight or in the early hours of the morning, depending on the location of the site.

In voice mail applications, I have found the busiest hours at the voice processing ports are when staff are least likely to be at their desks.(Makes sense, right?) Often that includes just before and just after the enterprise opens for business, when workers are either standing in line at the coffee machines, in meetings, or settling down for the day ahead; during the lunch break; just before and just after close of business; late at night; between 10 a.m. and 11 a.m.; and between 2:30 p.m. and 3:30 p.m. (usually the busiest period). Again, these busy periods naturally vary from enterprise to enterprise; you also have to factor in that IVR systems vary from application to application.

In south Florida, the lightning capital of America, during a thunder storm, calls disappear like a troop of baboons under a leopard attack. Calls return with a vengeance when the clouds clear. I expected the same phenomenon. But the storms do not last as long. While doing a needs synthesis or a huge shoe manufacturer, we found it easy to pin down the busiest hour of the month. After paychecks were handed out, callers would flock to the phones and dial up Accounting to clarify their deductions.

Trying to influence calling patterns also achieves mixed results. My travel agent client, Jan, advertises her special offers in the newspapers on Friday afternoons and Saturday mornings. In her advert, she advises people to call between 8 and 10 a.m. on Saturday morning. Do you think we could determine which hour or hours would be the busiest? Not so. The busiest hour was after 7 p.m. on Friday evening. People don't always read the whole advert, and others hope that they will catch an agent in the office "on the off-chance."

For the sake of this exercise, let's pretend that we received 1,000 calls in the hour.

### Determining the average call length

Don't despair, because the job gets easier. This part of the exercise is a little easier than forecasting the number of calls in the peak hour. Call center and telemarketing campaign planners work with scripts that require knowing the average length of a call. Inbound sales calls, say, those in response to an advert on TV, can be pretty uniform. Agents have scripts and software to help them complete a call within predetermined parameters.

Voice processing, automated attendant, IVR, and audiotext services are a little harder to determine the average call length for, but you at least have something to work with: the predetermined call paths and human-computer interaction. You know how long it takes to enter an extension number, or listen to voice mail, or enter a password, so you can do a little caller simulation.

Let's take computer telephonist and automated attendant services as an example. First, we list the most common types of callers in groups. These might be the following:

1 = Calls abandoned at the company greeting
2 = Calls abandoned after dialing an extension and receiving voice mail
3 = Calls that go through filtering and routing menus
4 = Calls that access the company directory
5 = Calls in which you leave a voice message
6 = Calls to retrieve voice mail (including message delivery)

Now we need to apply an average call length to each group. This task is the easy part. You'll need a stop watch or regular watch with a second hand (and a little time). Call the CT system and time each call group to arrive at its average call length.

At this point, you may ask yourself, how this can be done for a company that has no computer telephony system? The answer is that the consultant or vendor will usually do this with you; and they do this with a demo system or during the trial or evaluation period before you buy the hardware. The following list represents the average call length, measured in seconds, for each type of call:

1 = 5         4 = 300
2 = 4         5 = 200
3 = 300       6 = 300

Now you need to forecast the number of callers that in each group will have. Some estimates will be wild guesses, other estimates will be more educated. Sometimes there's just no way to predict what ends up happening. For example, I remember installing my first voice mail system and computer telephonist. The client was Comprehensive Property Services, and the systems were two of the first (PTT approved) systems ever installed in South Africa. After planning for weeks, we finally went live. On the first day, every caller abandoned the call at the greeting, much to my embarrassment. I had believed the callers would take to the system like fleas to dogs. (I never assumed how callers would react again.)

To illustrate a similar situation, in the example that follows, we deal with a typical installation in technology-savvy but often impatient New York City. Our sample size is 100 calls. The following list shows how that sample broke down into the call groups listed previously.

1 = 3 callers        4 = 5 callers
2 = 25 callers       5 = 40 callers
3 = 15 callers       6 = 12 callers

Now, follow these steps to figure out the parameters of the system that will meet our needs:

1. Determine the average call length. The following list shows the average call length for each group. From this list, we then total the results for each group and divide by the sample call size.

| Group | Calls | Length | Total (seconds) |
|---|---|---|---|
| 1 | 3 | 5 | 15 |
| 2 | 25 | 45 | 1,125 |
| 3 | 15 | 300 | 4,500 |
| 4 | 5 | 300 | 1,500 |
| 5 | 5 | 200 | 1,000 |
| 6 | 12 | 300 | 3,600 |

Total call duration = 11,740
Average call length = 117 seconds

2. Using these values, we can now determine the Erlang value:
      (1000 * 117) / 3,600 = 32.50 Erlangs

3. Next, determine the blocking tolerance level (people who don't connect). Because this is New York City, a tolerance level of 10 percent should be acceptable.

4. Determine the number of ports. Using the Erlang B Table again in Appendix B, a tolerance of 10 percent against an Erlang value of 32.36 (the closest value), indicates that we need a 32-port CT system. If the tolerance value was 1 percent, we would need 44 ports.

You will notice that in determining the call types, calls to retrieve voice mail are included in this groups. Some time back, when I knew less about CT than I do now, I estimated the number of inbound ports needed and then add four or eight ports for delivering messages. Using this method, frequently there was an extra board that stood idle all day. When you're measuring Erlangs, all your need to do is lump all message retrievals into one group and possibly increase the average length of this call, to compensate for tying up the port when dialing and performing call progress analysis (see Chapter 1 and Chapter 2).

A word of advice about the sample lists provided in the preceding example: Applications vary from site to site, city to city, and country to country, so you need to draw up lists that apply to your situation and conditions, not mine or someone else's. You may find that average call lengths double in your country due to factors that, for example, do not exist in the U.S. Here's a short list that elaborates on how factors can vary:

Rotary detection: Boy, does this task chew up ports. I hate rotary detection. As discussed earlier, the systems have to listen to the clicks (pulses per second) and work much harder to detect the digit than tone detection. Pulse-to-tone converters do not help much either. (It's easier to communicate with the Bushmen of the Kalahari Desert in Africa.)

Voice recognition: Voice recognition calls may very in length from person to person and nation to nation

Computer telephony hardware and software: Some systems on the market do not do such a good job of detecting touch tone, cutting though prompts, and integrating with the PBX. The hang-up detection factor is so important that it earns a section on its own further below. You should refer to it before finishing this section.

Telephony conditions in the external environment: One factor is noisy telephone circuits (which lowers the success rate of detecting touch tones) There are many others, such as weather conditions and so on.

User's culture, behavior and acceptance: Some cultures may find that human-computer dialog still goes against their grain, and users will take a long time to accept and work with voice mail features. In situations where most of the staff has never used voice mail, the average call length might be tenfold that of the company across the street.

## FCC

The FCC is the U.S. government agency that regulates telecommunications.

## Full-Duplex

Full duplex is simultaneous, bi-directional communication.

## Gain

The volume level of voice files in computer telephony systems.

## Glare

Glare is a condition that occurs when both the CPE and CO equipment try to make an outbound call on a line at the same time.

## Hang-up Detection

Many computer telephony buyers get lost in the jungle of features and options and seldom consider hang-up detection. Hang-up detection is one of the most important factors to be aware of when developing computer telephony systems and Windows telephony applications. The rule of thumb to follow is that a computer telephony system, upon getting a signal from the PBX or the exchange (either as loop current drop or a hang-up signal or a digital notification), must reset the line to provide access to another call within five seconds. Anything longer will push your port requirements through the roof and squeeze the wallet of your customer or client.

I am fanatical about hang-up detection. When I design computer telephony software, I make sure that we really put a lot of effort into this feature. In countries like South Africa, the telephone company typically does not provide you with a loop current off/on transition (a drop). In this case, the CT system relies on a hang-up signal.

At first, a basic algorithm using the Rhetorex DSP cards and its patented AccuCall(tm) technology had us resetting lines in under five seconds. Eventually we got it down to three seconds. But that time was still not good enough when compared to the rapid reset that you can achieve in some parts of the U.S. when you get transition in the loop current status. Eventually one of our smart engineers came up with a formula to detect the hang-up signal and reset the line in under a second. Often, it took less time to detect the signal, analyze it, confirm it as hang-up, and reset the line.

When upgrading a blocked 16-port system with the new "blitz-reset" software, however, we were stunned to see the port requirements drop to 12 ports. During the slower periods, port requirements further dropped to eight ports. The real advantage was in the financial area: This decrease could represent between a $5,000 and $10,000 savings in hardware and software license fees. The savings could translate even more with custom-made IVR software.

Hang-up detection and line reset should be an instant and automatic function of your computer telephony system. An exchange or PBX that provides loop current drop when a call is abandoned can make this capability happen. Often, as mentioned in Chapter 2, you don't get a loop current transition or other signal; just silence. The software then has to switch to a time-out function, which cannot be set to less time than it takes for two humans to enjoy some silence in their conversation.

Keeping that fact in mind, this situation is a less-than-happy one. The TAPI CTI-link changes things for the better because the PBX can alert the telephony server at the exact moment when a caller abandons the call.

Never consider a requirement to have the caller enter a terminating digit in order to reset the port. I have never encountered any person, and I don't think I ever will, who has the time to press the star digit or pound sign at the end of a call. When a caller is done, he or she just drops the line.

I recommend to end users that the best place to look for evidence of a user-affected terminating digit is the flow-chart documentation. "Go through the documentation of the proposed system and check for such a requirement. If it exists, make sure that this form of hang-up detection is not the only type employed. A time-out function should be somewhere in the picture. And preferably hang-up detection should be employed via signal analysis. If the two latter features are absent. Drop the system like a hot potatoe."

## Hook State

Hook state is the state of the telephone line (closed or open) at the CPE

## Hookswitch

The switch that connects or disconnects the device from the phone line. On a telephone, for example, this is the switch that is automatically activated when a user lifts the receiver from the cradle to get a new dial tone

## Hot Phone

A telephone whose connection is configured so that no dialing is required. As soon as the hot phone goes offhook, a destination phone (at a predetermined address) automatically begins to ring.

## Hunt/Hunt Group

When you order more than one telephone line from your local provider, an inbound call needs to know which line to call on if the first line is busy or in use. This searching process continues until a free line is found. This process is known as a hunt. If the first line is busy, the call will try a second line; if that's busy, it hunts for a third. The hunt group is a collection of lines that are organized in a fashion required by the enterprise. Hunting and hunt group management is important in computer telephony because you need to manage and flow the inbound calls to the computer telephony ports. Hunt groups can collapse; although you might think that business is quiet because several ports on the computer telephony system are inactive, the situation is really one in which several lines in the hunt group are lost.

## Inband

Transmitted within the data stream. Examples: POTS uses DTMF for inband dialing instructions and tones for inband notification that the remote station is busy or alerting.

## Intranet

Enterprise information network (LAN and WANs) technology blended with Internet technology, such as World Wide Web browsers.

## ISA

ISA means Industry Standard Architecture. It is the standard of IBM's PC bus architecture.

## ISDN

Integrated Services Digital Network, a set of standards for a new class of telephone services. ISDN is an entirely digital telephone service that can be installed by the local telephone company to replace the old analog local loop (the connection to the telephone company's nearest central switching office, or CO) with a digital line. Because long distance lines are usually digital already, replacing the local loop with an ISDN line provides "end-to-end" digital service.

For Telephony, ISDN's major significance is its ability to provide multiple channels on a single line.

## ISO

The International Standard Organization which sets standards for international data communications.

## Isochronous Transmission

The term isochronous comes from the Greek words "iso" for equal and "chronous" for time, meaning equal time (like real-time). Applied to data transmission isochronous transmission describes the transmission of time-sensitive data. Voice and video transmission is isochronous because the bi-directional communication needs to happen in real-time to be as natural as possible. When we talk over the telephone we usually do not notice any delays in transmission (ten or twenty years ago, and still in many places around the world, you can still experience a delay in receiving a return message from the other party on the line). When bi-directional data trasmission does not have to occur in real time or "at the same time" asynchronous transmission can suffice. (See asynchronous transmission above.)

## Key System

A switching system in which the phones have multiple buttons that permit the user to directly select incoming or outgoing lines. Key systems can typically support fewer users than PBXs, and their features are more limited.

## Kernel Mode Processing (Windows NT)

The Windows NT operating system model draws heavily from the client/server process model. The OS is divided (basically) into two components a user mode (client) and a kernel mode (server). Applications are mostly written to operate in user mode. When these application require services from the OS (like accessing the hard-disk, they make that request to the system services which operate in kernel mode. The kernel mode processes handle these request and serve functionality and information back to the clients running in user mode. The client/server design in Windows NT ensures stability of the OS by preventing processing applications from accessing devices directly.

TAPI 2.X has extension functions that allow you to create services that run in the kernel mode. These applications would not typically have user interface and, like device drives, they "sleep" in the background always ready to service requests from users.

## KSU

Key Service Unit refers to the switching electronics of a PBX or Key system.

## Line

The line is the electrical path between the CO and CPE.

## Line (TAPI)

One or more communication channels (accessed together as a unit) used by the application in performing telephony functions through TAPI and TSPI. A computer may provide its applications with access to multiple lines, and each line may provide different capabilities.

It is the choice of the service provider for a line to decide how to model its resources.

Note that a line need not correspond to a physical connection from the switch to the computer. For example, the realization of a line may involve a LAN-based server and a shared control link to the switch.

In essence, a line is any device that implements the line behavior defined by TAPI and by the SPI as the set of functions and messages for lines.

## Linear Data

Linear data refers to voice data that has been recorded at high resolution.

## Line Status

The status of the line at the CO.

## Loop Current

Loop current is the current that flows through the analog line from the telephone switch when the telephone is off hook. Loop current on the digital line means that the CO is off hook.

## Loop Drop

Loop drop is the transition from loop current on to loop current off.

## Loop Reversal

Loop reversal refers to the reversal of current flow to provide signaling information to the CPE.

## Loop Start

Loop start is a telephone protocol (see Chapter 1).

## Media

The media is whatever takes place on a line, usually on a 3.1-kHz audio bearer channel.

## Media Mode (TAPI)

A call's media mode describes what type of information the call is carrying, such as data or voice. An application can tell what media mode is indicated on an inbound call by examining a field in the call-info record. It can use this information, for example, to route the call to a more appropriate application, such as a data application for an incoming data call.

## Media Stream (TAPI)

The information carried on a call; that is, what actually is transmitted and received over the line, and can, with the necessary hardware, be read and written by a media stream API.

## MF

Multi Frequency. MF is a type of tone consisting of the ten tones that can be generated by the keypad.

## Micro Channel

Micro Channel is the proprietary bus developed by IBM for its PS/2 family of computers.

## m-Law

The PCM coding, compression, and expansion standard used in Japan and North America.

## MVIP

MVIP refers to a communications standard (hardware and software) that allows the printed circuit boards from different vendors or manufacturers to communicate with each other and exchange information between them in a PC. The boards communicate over a ribbon bus the connects one card to another each card. The MVIP standard was defined by Natural MicroSystems and Mitel with significant input from voice processing companies like Rhetorex. Several hundred companies support the MVIP standard now and they represent voice processing cards, switching cards, voice recognition cards, and fax cards.

## Off Hook

The state of a closed telephone line, which means that current is flowing.

## On/Off Hook Transition

The transition from on hook to off hook status.

## Out Of Band

Transmitted over a separate signaling channel. For example, with the media stream on the B channel, ISDN uses protocol messages on the D channel to indicate call states such as dialtone, ringback, and busy, and for signaling dialing instructions to the switch.

## PBX

Private Branch Exchange. A digital switch on the customer's premises that provides switching (including a full set of switching features) for an office or campus. PBXs often use proprietary digital-line protocols, although some are analog based. The user features provided by the different PBX vendors are generally similar.

## PABX

A PBX (private branch exchange) that is automated.

## PCM

Pulse Code Modulation. This is a method of compressing data in which a signal is sampled, and the magnitude of each sample is quantitized and converted into digital code.

## PCPM

Programmable Call Progress Monitoring (see Chapters 1 and 2). This refers to the capability of a computer telephony system to monitor the progress of a call under software control. The tones received during PCPM are matched against a table of tones loaded into memory during the normal functioning of a computer telephony system.

## Phone (TAPI)

A device that behaves as a telephone set. This is usually, although not necessarily, the phone already on the user's desk located "next to" the computer. Phones need not be physically connected to a computer, since other equipment, such as a LAN-based server with appropriate access to the switch may provide this logical connection. Note that TAPI and the Telephony SPI treat the phone and the line as separate devices that can be independently controlled by an application.

In essence, a phone is any device that implements the phone behavior defined by TAPI and by the SPI as the set of functions and messages for phones.

## Port

In computer telephony and telephony, a port is where you plug in devices (see Erlang above).

## POTS

POTS means Plain Old Telephone Service. Basic single-line telephone service for the public switched telephone network (PSTN). With some exceptions, POTS only supports making and receiving calls, and POTS lines can handle only one conversation at a time. To use a conventional modem and a telephone at the same time on a POTS system, two lines are needed.

## PPS

Pulses Per Second. The number of pulses transmitted or received in a second (ranges mostly from 9 to 22 in various parts of the world).

## PRI-ISDN

A "Primary-Rate Interface" ISDN connection, which in the U.S., Canada, and Japan consists of 23 64-Kbps B channels and one 64-Kbps D channel (23B+D). In Europe, PRI provides for 30 B channels and two D channels (30B+2D).

## PVC

Positive Voice Control. An algorithm (Rhetorex, Inc.) that distinguishes between voice, data, call progress tones, noise, and silence.

## PSTN

Public Switched Telephone Network.

## Ringback

The tone heard by a calling party when, at the called-party's end, the telephone is ringing or the system is otherwise being alerted of the incoming call (see Chapter 1 and Chapter 2).

## SCSA

Signal Computing System Architecture (SCSA). This is a comprehensive open architecture for providing multiple user computer telephony services in a client server environment. An SCSA server offers call control capabilities, such as the capability to make calls, answer calls, route calls, monitor calls, conference calls, and manipulate information (affect the content of calls) over the telephone network. SCSA servers also offer the capability to process various types of media (for example voice, fax, text, email) and make conversions between those types of media in order to present the information in a more useful form.
SCSA is composed of a hardware model and a software model. The SCSA Hardware Model defines the interfaces and protocols for a real-time communication bus and a flexible, high capacity, distributed switching fabric.

The SCSA Telephony Application Objects Framework(tm) (TAO) defines a standard set of software interfaces, protocols, and services that constitute the software model. These interfaces are open, object oriented, and hardware independent.

The SCSA TAO Framework contains a standard set of APIs that lets multiple client applications share a common server and all of the call control and media processing resources within the server. Below the API level, the SCSA TAO Framework consists of a service provider framework that includes a standard set of services that manage the server without requiring any application involvement. It also consists of a standard protocol that allows for communications within the server.

Together, these models are designed to offer tight interoperability and extensive functionality, not only for today's computer telephony systems, but also for tomorrow's more complex systems and new technology (See TAO below).

## SIR

Speaker independent recognition. SIR refers to the recognition of a vocabulary by a computer telephony system, albeit limited. The speaker can be any caller to the system (see also Voice Recognition below).

## Speech

Speech is human speech, a specific type of voice. Telephone networks treats speech and voice differently, because speech can be modeled and compressed more than voice. Voice is less likely to be compressed than speech (especially on international calls), because compression can interfere with some high-speed voice band data and fax transmissions. In contrast, speech can be understood when re-expanded even after being compressed to 1/18 its original bandwidth.

## Spike

A temporary and unexpected signal on the telephone line.

## Station

The station (in computer telephony) is the communications equipment at the end of the telephone line. It also refers to the physical connection or integration of this equipment with the PC. The PC as a communications device thus comprises the station.

## Switch

Telephone switch. A piece of equipment capable of establishing telephone calls. Within the context of TAPI and TSPI, a switch can be a PBX, a key system, or a central office.

## Synchronous Completion (Win32 API)

An application calls a functions in the DLL, which then completes and returns to the application. If it completes immediately, this completion is known as synchronous, but if it is sent off to another system entity and the application goes on to other activities before the function completes (and the system later sends a message to the application announcing the function's completion), that completion is known as asynchronous.

## Talk Off

Related to cut-through, talk-off occurs when the human voice produces what sounds like a tone-tone digit. This can happen during a recording; the system will recognize the tone and terminate the event.

## TAO

TAO stands for Telephony Applications Objects framework. It is currently maintained and supported by the Enterprise Computer Telephony Forum (ECTF). (See both ECTF and SCSA entries above.)

TAO first appeared in 1994 as the computer telephony industry's first comprehensive, fully distributed voice processing API. It was originally called the SCSA API, but failed to obtain multi-vendor support. Later, it was incorporated into the ECTF S.100 specification. In 1995 it was formally incorporated in the ECTF API.

## Key TAO Feautures:

- Supports both MVIP and SCSA hardware components by design
- It is an object oriented technology and describes objects to play, record, detect DTMF, generate DTMF, perform text to speech and voice recognition and more

- It establishes the concepts of resources, resource groups, and pools
VIt includes application profiles that define resources required by each application
- It provides a Call Router system (SCR) in which calls are delivered already bundled with the resources it will need
- Calls may be passed between applications on the same machine as well as to other machines
- It deploys a call control model that makes use of other standard call control models such as TAPI and TSAPI (the Novell, AT&T et al telephony API).

Despite its promise the TAO are still being designed. It is also a complex API for service providers and application developers mainly because of its huge scope. It also competes directly with Microsoft's media APIs such as MS Fax, and the Speech API (the Speech API from Microsoft includes Text-to-Speech, Speech-to-Text, Voice Command and Voice Recognition functionality).

## TAPI

The Windows Telephony API (TAPI) defines the interface that applications use to access telephony functions in Windows. The API is a collection of C language function definitions, message definitions, type and data-structure definitions, along with descriptions of their meanings in English. TAPI is a joint effort between Intel Corporation and Microsoft Corporation, started in 1992 to provide a standard way to integrate the telephone with the personal computer. It was officially released on May 4, 1993.

**Here is a reprint from that historical Press Release**

"Future products to add telephony functions to the PC will be based on a new specification, called Windows Telephony Application Programming Interface (TAPI), that Intel, Microsoft, and industry participants have co-developed for the Microsoft Windows operating system. The specification was reviewed and is supported by approximately 40 companies, including major telephone switch manufacturers, PC and peripheral manufacturers, software developers and network providers

**Products to Add New Functions to PCs**

Products based on the specification will enhance existing PC applications and enable new applications. Applications such as database managers, personal information managers, spreadsheets, and word processing will benefit by gaining direct access to the phone network. New communications applications enabled by Windows Telephony include:

- Visual call control that uses the PC's graphical user interface for call fowarding, conferencing and call transfer
- Integration of electronic mail, voice mail and fax
- Desktop audio and video conferencing
- Wide area networking that allows PCs to use the telephone network for both voice and data transmissions

The specification is intended to insulate PC users and applications developers from the underlying computer hardware, connection model or telephone network being used, including PBX, ISDN, Centrex, cellular or analog telephone service.

"We at Lotus see the Windows Telephony API as the first of a suite of application programming interfaces we can use to incorporate telephony into our Workgroup Computing products," said Alex Morrow, general manager of Cross Product Architecture at Lotus Development Corporation. "By making good use of the API in our workgroup applications, we're hoping to streamline business communications."

"We fully support the Microsoft and Intel initiative on Windows Telephony for desktop applications," said Tom Lowery, vice president, Multimedia Applications, Northern Telecom. "Open architectures are essential for stimulating the introduction of high-value applications that require seamless, transparent interworking between desktops." Northern Telecom will support Windows Telephony as part of the VISIT Access program, which is the open architecture of the VISIT Multimedia product line.

"The specification will allow us to help our customers achieve the efficiencies that come from integrating telephony and PCs," said Peter Pribilla, group president, Siemens Private Communication Systems Group and president and CEO of ROLM. "This alliance is another substantial step in our effort to provide high-efficiency desktop solutions."

**Extending the PC Architecture for Real-Time Communications**

"We are pleased to see such strong, unprecedented support from the software, PC and telecom industry for extending the PC architecture," said Ron Whittier, Intel vice president and general manager of Intel's Architecture Software Technology Group. "As Intel introduces products based on this work later this year, we are looking forward to working with our partners in the software and telecommunications industry to bring exciting applications to PC users."

"Windows Telephony allows the integration of information on the PC with real-time communications, which brings us another step toward the vision of 'information at your fingertips,'" commented Jonathan Lazarus, vice president, Systems Strategy at Microsoft. "The telephone has long been an office staple upon which employees have relied to be effective in their jobs. With Windows Telephony, resulting products will not only combine the strengths of the telephone with the power of the computer, they will open up a whole new world of applications."

The Windows Telephony application programming interface specification is part of the Microsoft Windows Open Services Architecture (WOSA), which provides a single set of open-ended interfaces to enterprise computing services. WOSA encompasses a number of APIs, providing applications and corporate developers with an open set of APIs to which applications can be written and accessed. Feedback received through the WOSA open process assures a fully integrated framework and a complete set of telephony features for software developers, telecommunications manufacturers, and personal computer vendors to deliver products. WOSA also includes services for data access, messaging, software licensing, connectivity and financial services.

Today, Windows Telephony is focused on enabling the desktop and will be extended to server environments in future releases. Additionally, Intel and Microsoft will encourage open cooperation with other system software providers to bring the functionality provided by the specification to other computing platforms."

## TAPI DLL

The client software module that interacts directly with the Telephony service provider by means of the Telephony SPI. The TAPI DLL exports the Windows Telephony API to its clients. Its clients in turn are usually applications, which may be DLLs operating on behalf of applications.

## Telematics

Essentially information exchange. It refers to information transmission services such as fax and telex.

## Telephone Network Interface

The Telephone Network interface enables computer telephony systems to communicate with specific telephone networks. Calls arriving from telephone networks can be carried on a variety of lines, from analog loop start and DID (DDI) lines to digital T-1, E-1, and primary rate Integrated Services Digital Network (ISDN) lines. Network interfaces also interpret signaling coming across the telephone line, provide data buffering, and include surge protection circuitry.

## Telephonist

A species of human (homo telephono) who have evolved from the now highly endangered species that only handled the answering, routing process and message taking functions of the enterprise-wide telephone service. Since the advent of the voice processing board and the era of computer telephony, the duties of the telephonist include business administration, word and data processing, telephone management, call accounting, PBX maintenance and management, user training, public relations, Press liaison, computer telephony scripting, voice recording, data back-up and information archiving, executive assistance, and much more. The earlier species that did nothing more than answering, routing and message taking will likely be extinct by the end of the decade. Although there are very few left in the world, conservation authorities have no inclination to preserve this animal.

## Terminating Tone

A call progress tone that indicates the termination of a connection. Terminating tones are received with remote CPE or callers end the communication.

## Text-to-Speech

Also known as TTS. It is the technology for converting text (ASCII) into a synthetic speech. It is used in computer telephony, especially interactive voice response, to read the values and information in database tables to callers. TTS is an economical way of giving customers telephone access to information that would be too expensive or impractical to record using voice technology.
TTS is also used to read text in a direct human-computer dialog. A computer can, for example, read a book to a blind person; it is also used to help people who cannot speak.

## Tone Detection

Tone detection and processing includes the capability to receive, recognize, and generate specific telephone and network tones. This capability allows a computer telephony application to place a call and monitor its progress. Tones that are processed include: busy tones, ringing, dial tones, fax machine tones, modem tones, and MF tones.

## Universal serial bus (USB)

A new bus technology moving to the center of the CT and PC telephony industry is Universal Serial Bus (USB). Among a number of objectives USB makes it easier to connect a telephone to a computer. It is rapidly being adopted all the major computer industry players and is now fast becoming a standard.
Here is how it works:

One of the limiting factors of the new era in information technology, especially the breakneck developments in the PC industry, has been the dearth of peripheral devices that can be attached to the computer. These devices include modems, keyboards, mice, printers, scanners, joy-sticks video cameras, microphones, speakers and more.

To connect these devices to the computer so that data can move to and from the device, you have to insert adapter cards into the interface slots on the computer's motherboard. A motherboard typically has eight slots. With so many interface cards being manufactured, such as voice processing boards, the internal peripherals have been fighting for slot space.

The PC industry is also trying to squeeze all the PC innards into smaller cases so that notebook computers can keep getting smaller while becoming more powerful (imagine a full-featured PBX and voice mail systems inside a little notebook; watch your computer store). Now what about trying to connect the telephone to the PC.

USB is emerging as a computer telephony standard for connecting the telephone to the computer. It employs Plug and Play capability and hot attach/detach (hot docking) technology that empower you to configure your computer according to the wildest inventions of your imagination. You can attach and remove a device from the computer at any time without having to power down or reboot the computer. And you can connect all kinds of devices to the computer wherever you can locate a USB port.

For example, if a port is located on the monitor, then that's where you can plug in a device such as a headset. If there's a port on your keyboard, then you can attach your headset there. You can also detach your headset and attach it to another computer without interrupting the computing environment. Because of the functionality that USB provides, companies will soon be installing USB hubs on their LANs.

**Here are some of the important specifications of USB:**

- USB supports up to 127 devices, including gaming devices and virtual reality goggles
- It has a 12Mbps design and can transfer data isochronously and asychronously
- It supports up to 5 meters of cable
- It supports daisy chaining through a tiered, star, multidrop topology

Why is USB so important for CT? For starters, you can create a CTI link without specialized add-in cards by using USB. Second, this bus supports high-speed digital interfaces like ISDN, T1, and E1 (T1 and E1 are the North American and European digital standards, respectively. Third, you can unplug telephony devices from your desktop computers and connect them to your portables, further eliminating the cloning syndrome.

## Windows Telephony Application

Any software that uses the Windows Telephony API or TAPI. The term "application" is used in its broadest sense possible; it need not necessarily be a user-level program, but can also be a dynamic link library (DLL) or system function that uses the API and provides higher level services by means of its interface. TAPI 2.X has functionality, for example, that allows you to create services that run in the background as non-user interface processes.

## Windows Telephony Service Provider

The conglomerate of software code (DLLs, device drivers, firmware) and hardware (add-on hardware, server, phone set, switch, network) that jointly implement the Telephony SPI.

## Windows Telephony SPI

The Windows Telephony SPI (Service Provider Interface) is the interface that a service provider must implement to make its telephony services available to applications through the API. The SPI is a collection of C-language function definitions, message definitions, type and data-structure definitions, along with descriptions of their meanings in English.

## Voice

Anything that can be transferred on a POTS network, namely any signal that fits on a 3.1 kHz-bandwidth channel. Voice can consist of voiceband-modulated data or facsimile signals or human speech.

## Voice Processing

Voice is the fundamental technology at the core of most Computer Telephony systems. It encompasses both the processing and the manipulation of audio signals in a Computer Telephony system. Tasks include filtering, analyzing, recording, digitizing, compressing, storing, expanding, and replaying signals.

## Voice Recognition

Voice recognition is the technology that enables a computer telephony to recognize and understand the human voice. Voice recognition systems require training and are becoming outmoded in favor of speech recognition technology in which any speaker can have a conversation with a computer (see ASR).

## VoiceView™

VoiceView, a protocol developed by Radish Communications, enables switching between voice and visual data in very rapid sequence during the same call. It is supported by Microsoft Windows 95.

# Appendix B
# Computer Telephony OLE Component (OCX) Example

The Visual Voice voice control implemented in Object Pascal (Delphi Version 2). The following is an example of a high level telephony component which further hides the complexity of APIs such as TAPI. Under the 16-bit operating systems (Windows 3.XX) this approach to computer telephony application development was steeped in controversy as to its ability to stand up to mission critical applications. The controversy still rages (not only for telephony), just read the threads in the USENET and you will see what I mean. Many still believe you can't code any high-end robust computer telephony application by setting properties and events in high-level components. But you will see in this code that it is even possible to create not only true multi-tasking applications, but multi-threaded ones at that.

This example of a computer telephony component used in a multi-line order information system was supplied courtesy of the Stylus Innovation division of Artisoft, Inc.

```
unit multifrm;

interface

uses
   SysUtils, Windows, Messages,Classes,Graphics,Controls,
   StdCtrls, Forms, DBCtrls, DB, ExtCtrls,Mask,DBTables,
   Dialogs,OleCtrls, VV, Voice, ComCtrls;

type
   TLineInfo = record        {Information for the
                              current line}
```

```delphi
    VoiceCtrl : TVoice;          {The Visual Voice voice
                                    wrapper in Delphi}
    Data :      TDataSource;     {The Delphi
                                    Database Link}
    Table :     TTable;          {The Delphi Table
                                    Component}
    Image :     TListItem;       {Item in the ListView
                                    component}
end;

TVoiceThread = class(TThread)    { Our new thread class
                                    in which the control
                                    will be executed}
public
   Line : TLineInfo;             {Information about
                                    the line}
   procedure Execute; override;
   constructor Create(L: TLineInfo);
end;

TOrderForm = class(TForm)
   ListView1: TListView;
   procedure StartUP(Lines : Integer);
   procedure VoiceCtrlRingDetected(Sender: TObject);
   procedure VoiceCtrlLineDropped(Sender: TObject);
   procedure FormDestroy(Sender: TObject);
   procedure FormResize(Sender: TObject);
   procedure FormShow(Sender: TObject);
  private
   { private declarations }
  public
   { public declarations }
  end;

var
  OrderForm: TOrderForm;         {The form: GUI }
  LineInfo:  Array[1..24]
     of TLineInfo;               {Line information (up to
                                    24 lines)}
  LineCount: Integer;            {How many lines?}

Implementation //In Delphi this where the fun starts
```

```
uses intro;

{$R *.DFM}  //Compiler directives

constructor TVoiceThread.Create(L: TLineInfo);
begin
    inherited Create(False);
    FreeOnTerminate := True;     {Automatically  kill the
                                  thread when done}
    Line := L;
    // Open the database
    Line.Table.Active := True;   {Open the database. Note
                                  that this  can be set
                                  at both runtime an
                                  design time.}
end;

procedure TVoiceThread.Execute;

var
   Ordernum : String;

begin
// Start try-except block
With Line Do
Begin
    Try
        // Pick up
        Image.ImageIndex := 1;
        Image.Caption := 'Picking up';
        VoiceCtrl.Pickup(Null);

        // Play 'Hello.  Welcome to the Order Status
              System.'  [Greeting]
        Image.Caption := 'Playing';
        VoiceCtrl.PlayFile('..\greetin1.vox', Null, '',
                            Null, Null);

        // Query for GetDigits MaxDigits=5 [GetOrderNum]
        VoiceCtrl.PlayFile('..\getorde1.vox', Null,
                            Null, Null, Null);
        Image.Caption := 'Getting digits';
```

```
        OrderNum := VoiceCtrl.GetDigits(5, Null, Null,
                                        Null);

        // Search for OrderNum in the OrderTable

        Image.Caption := 'Searching';
        If Table.FindKey([OrderNum]) then
          Begin
                // 'Order number...' 1234 'was shipped
                    on...' Wed Mar 15
                // 'and totalled...' 12.34   [Play Order]
                    (This is voice over the line)
            Image.Caption := 'Playing';

            VoiceCtrl.PlayString('..\ordernu1.vox|F, ' +
            OrderNum + '|C, '   + '..\wasship1.vox|F, ' +
vvDateStr(Table.Fields[1].AsString) +
                                         '|D|ddd mmm d,
..\andtotal.vox|F, '+ Line.Table.Fields[3].AsString +
'|M', Null, Null, Null);
          End
        Else
          Begin
                // Play not found message
                Image.Caption := 'Playing';
                VoiceCtrl.PlayFile('..\isnotva1.vox',
                        Null, Null, Null, Null);
          End

        // Play 'Thank you for calling.  Goodbye.'
                          [Goodbye]
        Image.Caption := 'Playing';
        VoiceCtrl.PlayFile('..\goodby1.vox', Null, Null,
                          Null, Null);

        // Hang up
        Image.ImageIndex := 0;
        Image.Caption := 'Idle';
        VoiceCtrl.Hangup(Null);
    Except
        On VoiceError : Exception Do
```

```
              Begin
                 // Display an error box if it wasn't a
                     LineDropped error
                 If VoiceCtrl.ErrorNumber
                     vvpLineDropped Then
                   MessageDlg(VoiceError.Messag
                       mtError, [mbOK], 0);

                 // Make sure we are stopped and hung up
                 VoiceCtrl.Stop();
                 VoiceCtrl.Hangup(Null);
                 Image.ImageIndex := 0;
                 Image.Caption := 'Idle';
              End;
      end;
end;
end;

procedure TOrderForm.StartUP(Lines : Integer);
var
   I :         Integer;
   Item :      TListItem;
   Images :    TImageList;
   Icon :      TIcon;
   Dir :       String;

begin
     // Check the passed parameter for how many lines
         to start

   If Lines = 0 Then
      LineCount := 2
   Else
      LineCount := Lines;

     Application.Title := 'Order Status ' +
IntToStr(LineCount) + ' Lines';

   // Create an image list and fill it with the two
images we need
```

WINDOWS TELEPHONY

```
Images := TImageList.Create(Self);
Icon := TIcon.Create;
Icon.LoadFromFile('handon.ico');
Images.AddIcon(Icon);
Icon.LoadFromFile('handoff.ico')
Images.AddIcon(Icon);

   // Tell the ListView control which images to use

   ListView1.LargeImages := Images;

   // Initialize all the lines, we have to start at
        the top to keep from
   // conflicting with the new controls that autoa
        locate to line 1

   For I := LineCount downto 1 Do
       Begin
            frmIntro.lblCurrent.Caption :=
                IntToStr((LineCount - I) + 1);
            frmIntro.lblCurrent.Refresh;

            // Create the voice control, set its
                 phone line, and register the events
            LineInfo[I].VoiceCtrl :=
                 TVoice.Create(Self);
            LineInfo[I].VoiceCtrl.DeallocateLine;
            LineInfo[I].VoiceCtrl.AllocateLine(I,
                 Null);
            LineInfo[I].VoiceCtrl.OnRingDetected :=
                 VoiceCtrlRingDetected;
            LineInfo[I].VoiceCtrl.OnLineDropped :=
                 VoiceCtrlLineDropped;

            // Create a data control and a table
                 control for this line

            LineInfo[I].Data :=TDataSource.Create
                  (OrderForm);
            LineInfo[I].Table :=TTable.Create
                  (OrderForm);
```

```
            // Setup the table and database
            GetDir(0, Dir);
            LineInfo[I].Table.DatabaseName := Dir;
            LineInfo[I].Table.TableName := 'order.dbf';
            LineInfo[I].Table.IndexName := 'OrderNum';
            LineInfo[I].Data.DataSet :=
                LineInfo[I].Table;

            // Initialize the image in the ListView
               control

            Item := ListView1.Items.Add;
            Item.Caption := 'Idle';
            Item.ImageIndex := 0;
        End;

    frmIntro.Hide;
end;

procedure TOrderForm.VoiceCtrlRingDetected(Sender:
TObject);
var
    I : Integer;
begin
    // Set some last minute information on the phone
       line

    I := TVoice(Sender).PhoneLine;
    LineInfo[I].Image := ListView1.Items[I - 1];

    // Start up the thread with the info we need
    TVoiceThread.Create(LineInfo[I]);
end;

procedure TOrderForm.VoiceCtrlLineDropped(Sender:
TObject);
begin
    // Set the LineDropped error
    TVoice(Sender).SetError(vvpLineDropped);
end;

procedure TOrderForm.FormDestroy(Sender: TObject);
```

```
var
    I : Integer;
begin
      // Stop and Hangup all voice controls and free any
         components we created

      For I := LineCount downto 1 Do
          Begin
                LineInfo[I].VoiceCtrl.Stop();
                LineInfo[I].VoiceCtrl.Hangup(Null);
                LineInfo[I].VoiceCtrl.Free;
                LineInfo[I].Data.Free;
                LineInfo[I].Table.Free;
                LineInfo[I].Image.Free;
          End;
end;

procedure TOrderForm.FormResize(Sender: TObject);
begin
      // Make sure the list view is always the right
         size
      ListView1.Width := OrderForm.ClientWidth;
      ListView1.Height := OrderForm.ClientHeight;
end;

procedure TOrderForm.FormShow(Sender: TObject);
begin
      // Show the startup form
      frmIntro.Show;
end;

end.
```